T0213615

Fundamentals of Thermodynamics

Fundamentals of Thermodynamics

John H. S. Lee and K. Ramamurthi

CRC Press
Taylor & Francis Group
Boca Raton London New York

CRC Press is an imprint of the
Taylor & Francis Group, an **informa** business

First edition published 2022
by CRC Press
6000 Broken Sound Parkway NW, Suite 300, Boca Raton, FL 33487-2742

and by CRC Press
2 Park Square, Milton Park, Abingdon, Oxon, OX14 4RN

© 2022 John H. S. Lee and K. Ramamurthi

CRC Press is an imprint of Taylor & Francis Group, LLC

ISBN: 978-1-032-12312-7 (hbk)
ISBN: 978-1-032-12300-4 (pbk)
ISBN: 978-1-003-22404-4 (ebk)

DOI: 10.1201/9781003224044

Typeset in Times
by codeMantra

Contents

Symbols

A, a	Helmholtz function and specific Helmholtz function
atm.	atmospheric pressure
amu	atomic mass unit
b	number of atoms
c	specific heat
C_p, c_p	constant-pressure thermal capacity and constant-pressure specific heat
C_V, c_V	constant-volume thermal capacity and constant-volume specific heat
COP	coefficient of performance
d	exact differential
G	Gibbs function; Gibbs free energy
g	specific Gibbs free energy, degeneracy; gravitational field; gram
f	fugacity: number of degrees of freedom
f_0	fugacity in the limit of pressure tending to zero
H, h	enthalpy and specific enthalpy; Plank's constant
HE	heat engine
HP	heat pump
I	irreversibility
J	mechanical equivalent of heat
K	equilibrium constant
κ	Boltzmann constant
KE	kinetic energy
kg	kilogram
L	Lagrangian
M	molecular mass, number of atoms in a reaction
m	mass; meter; mass of particle
m_0	mass of standard particle
$\dot{m}$	mass flow rate
N	number of molecules/particles; number of species in a reaction
N_0	Avogadro's number
n	amount of moles; quantum number
n_i	number of moles of specie i
PE	potential energy
p	pressure
p_{cr}	critical pressure
p_i	partial pressure of component i in the mixture
p_r	reduced pressure
Q	heat transfer
$\dot{Q}$	rate of heat transfer
R	gas constant; reversible engine
R_0	universal gas constant
S, s	Entropy and specific entropy
T	absolute temperature, thermodynamic temperature

T_0	environment temperature
T_r	reduced temperature
U, u	internal energy and specific internal energy
V, v	volume and specific volume
V_i	partial volume of specie
v_r	reduced volume
$\tilde{v}$	specific molar volume
$\bar{v}$	velocity
W	work, permutation
W_S, W'	non-displacement work
$\dot{W}$	rate of work
W_R, W_{rev}	reversible work between two states
w	work per unit mass
X	irreversible engine; thermometer reading
x	mole fraction or concentration
y	mass fraction
Z	compressibility factor
z	partition function

Greek letters

α	pressure coefficient, Lagrangian multiplier
β	coefficient of volume expansion, Lagrangian multiplier
γ	specific heat ratio c_p/c_V; ratio of thermal capacity C_P/C_V
γ_x	stoichiometric coefficient of specie x
Δ, δ	increment
∂	partial differential
ε	energy of particle
η	efficiency; Joule coefficient
θ	empirical temperature
κ	degree of reaction
κ_S	isentropic expansion coefficient
κ_T	isothermal expansion coefficient
λ	Lagrangian multiplier; scale factor
μ	chemical potential
μ, μ_π	Joule Thomson coefficient
$v_{i,j}$	number of atoms of element j in specie i
Π	product
ρ	mass density
Σ	summation
τ	relaxation time, characteristic time
ϕ	availability; fugacity coefficient
ϕ^*	availability coefficient including work
χ	degree of reaction
ψ	open system availability

Subscripts

ad	adiabatic
atm	atmospheric
c, cr	property at critical point
env	environment
f	formation
HE	heat engine
HP	heat pump
i	i^{th} component, i^{th} state, ice, ideal
j	j^{th} component, phase, j^{th} state
max	maximum
p	constant pressure
R	reservoir, reversible engine
r, R	reduced property
R, rev	reversible
res	reservoir
s	isentropic: steam
sys	system
T	isothermal
tp	triple point
v	constant volume
x	space coordinate in X direction, irreversible process
y	space coordinate in Y direction
z	space coordinate in Z direction
0	environment

Superscripts

~	specific property per mole, ideal gas property
0	equilibrium state, isolated system, standard state
*	ideal gas; maximum work
•	rate

Preface

Thermodynamics is an essential subject in mechanical engineering and almost all universities have a one- or two-semester course on thermodynamics. Accordingly, there are numerous textbooks written on the subject. Thermodynamic texts have progressively increased in volume, and the present-day books are close to a thousand pages. The increase in volume is due to increase in the number of worked examples, homework problems, and various study aids such as computer software. The scope of the subject has also increased to include current topics on renewable energy, advanced propulsion, environmental issues, bioengineering, etc. Many thermodynamic texts include chapters on compressible flow for which a separate course is generally part of the curriculum.

The present book deals with the basic concepts of thermodynamics. These tend to be overlooked in the context of covering a wide range of topics. The fundamentals of thermodynamics are to a large extent governed by the first and second laws. The first law is essentially the conservation of energy, which is generally discussed in a first-level course in mechanics. The subtle contribution from the first law is in defining internal energy. The second law is the unique part of thermodynamics which is distinctly very much different in approach from the other subjects. The major portion of the book deals with the second law and its consequences.

The internal energy from the first law and entropy, reversible work, irreversibility and thermodynamic equilibrium from the second law are dealt with in the different chapters. The relationships among the thermodynamic functions, properties and coefficients are obtained and their utility demonstrated. The treatment is not restricted to ideal gases; rather the thermodynamic properties, functions and processes are derived in a general manner and those for ideal gases are obtained as limiting cases of the generalized results. The maximum work from a system is determined and the importance of reversible paths during change of state is discussed. The role of the environment in influencing entropy and work is specifically considered. Thermodynamic equilibrium is addressed through isolated systems and systems interacting through heat, work and mass exchange. Thermal, mechanical, chemical and phase equilibrium in simple systems are dealt with. Species formed at equilibrium in a chemical reaction are also considered. The molecular basis for the first law comprising internal energy, work and heat and the basis for the second law from reversibility and entropy is briefly discussed.

The book would supplement and be a source of reference to go with the standard textbooks that are in abundance. The book would, in particular, assist the undergraduate, post-graduate and research students as well as practicing engineers to appreciate the basics of thermodynamics and apply the concepts gainfully for solving issues in the areas of energy, power, propulsion and environment.

Authors

Professor John Lee is an Emeritus Professor of Mechanical Engineering at McGill University, Montreal. His scientific interests are in thermodynamics and fluid mechanics. He has been carrying out fundamental and applied research in combustion, detonation and shock-wave physics for the past 50 years. He has numerous scientific publications in the area. He is the author of the books *The Detonation Phenomenon* and *The Gas Dynamics of Explosions*.

Professor K. Ramamurthi was a Deputy Director at the Indian Space Research Organization and subsequently a Professor in the Mechanical Engineering Department at the Indian Institute of Technology Madras at Chennai. His notable contributions to research have been in instability phenomenon, rocket propulsion and blast wave mitigation. He is the author of the books *Rocket Propulsion* and *Modeling Explosions Blast Waves*.

1 Fundamental Concepts

1.1 SYSTEM AND ENVIRONMENT

Thermodynamics is a study of the interaction of a system with its environment. A system is part of the universe contained within a prescribed boundary that we deal with. Everything outside the boundary is called the environment. Thus, the system and its environment constitute the universe.

The boundary of a system may be a real physical boundary or could be imaginary. The boundary may be of arbitrary shape and could move when the volume within the boundary changes such as when the system expands or contracts when interacting with the environment.

An isolated system has a boundary that does not permit mass or energy exchange across it. For a closed system, the boundary permits only energy exchange. If mass as well as energy is exchanged across the boundary, the system is called an open system.

1.2 STATE OF A SYSTEM

The state of a system is defined by a set of measurable macroscopic parameters. The state can be measured only if the variables defining the state are invariant with respect to time and space within the system; that is, the system is in equilibrium. It is the equilibrium state of a system that is defined by the state variables of a system.

State variables that depend on the mass of the system are called extensive variables, for example, energy and volume. State variables that are independent of the mass are called intensive variables, for example, pressure, density and temperature.

The specific value of an extensive property is the extensive property divided by the amount of substance, for example, specific volume $v = \dfrac{V}{m}$, where V is the volume and m is the mass.

1.3 SIMPLE SYSTEMS

We shall be dealing mostly with simple systems. A simple system is one which is homogeneous, isotropic and chemically inert. It is sufficiently large in that surface effects can be neglected. In other words, we may define its energy without considering the surface energy due to the boundary separating the system from the environment. The external forces arising from electromagnetic, gravitational and similar environmental effects are also not considered in contributing to the energy of the simple system. So the simple system can generally be defined solely by its energy U, volume V and amount of mass in the system, that is, (U, V, m_i) where m_i is the mass of the different chemical components "i" in the system.

DOI: 10.1201/9781003224044-1

1.4 MASS, MOLECULAR MASS AND MOLES IN A SYSTEM

The mass of a system is the number of molecules N contained in the system multiplied by the mass of each of the molecules in it. Since the mass of a molecule is very small, it is measured in terms of the mass of a standard particle that is chosen to have a mass one-twelfth the mass of an isotope of carbon $\overset{12}{C}$. The mass of the standard particle m_0, known as the atomic mass unit (a.m.u.), is 1.661×10^{-24} g.

The mass of a molecule of a substance is therefore expressed in units of the standard atomic mass unit m_0, namely, mass of the molecule m divided by m_0, that is, $M = \dfrac{m}{m_0}$. M is called the molecular mass.

As an example, the molecular mass of a hydrogen molecule is given as $M_{H_2} = \dfrac{m_{H_2}}{m_0}$, where m_{H_2} is the mass of the hydrogen molecule and m_0 is the mass of the standard particle. It is also spoken of as molecular weight since almost all experiments are carried out in the vicinity of the Earth's surface where the gravitational constant is the same. We will use the words molecular mass and molecular weight without differentiating between them.

The number of molecules in a macroscopic system is, in general, very large, and we therefore measure it in the unit of mole. A mole is defined as the number of standard particles N_0 in 1 g of it, that is, $N_0 = \dfrac{1}{m_0} = \dfrac{1}{1.661 \times 10^{-24}} = 6.023 \times 10^{23}$. N_0 is called Avogadro number.

The number of the moles of a substance comprising of N molecules is $n = \dfrac{N}{N_0}$.

We can write the molecular mass as

$$M = \frac{m}{m_0} = \frac{m}{m_0} \frac{N_0}{N_0} = mN_0 \tag{1.1}$$

since $m_0 N_0 = 1$ g.

The molecular mass M therefore equals mN_0 in unit of grams and is the mass of 1 mole of the substance in grams. Thus, 1 mole of hydrogen has a mass equal to 2 g, and 1 mole of nitrogen is 28 g and so on. Similarly, the number of moles n of a substance of mass m g is m/M.

For a mixture of gases consisting of N different constituents, the mole fraction of the ith constituent in it is

$$x_i = \frac{n_i}{\displaystyle\sum_{i=1}^{i=N} n_i}$$

where n_i mole is the number of moles of the ith constituent in it and $\displaystyle\sum_{i=1}^{i=N} n_i$ is the total number of moles in the mixture.

Similarly, if the mass of the ith constituent is m_i, the mass concentration of the ith constituent is

$$y_i = \frac{m_i}{\displaystyle\sum_{i=1}^{i=N} m_i}$$

The sum of the mole fractions and mass fractions is unity, namely,

$$\sum_{i=1}^{N} x_i = 1, \sum_{i=1}^{N} y_i = 1$$

1.5 INTENSIVE VARIABLES DEFINING A SYSTEM

Energy U, volume V and mass m or equivalently the moles n, which define a system, are based on the extent of a system. The energy per unit mass $u = \dfrac{U}{m}$ and the specific volume $v = \dfrac{V}{m}$ are independent of the extent and are intensive variables. In the following, we define the intensive variables pressure and temperature for defining a simple system. These are independent of its extent and are the so-called intensive variables.

1.5.1 PRESSURE

The pressure p is the force per unit area and acts normal to the surface and is independent of the orientation of the surface. The unit of pressure is Newton per square meter (force per unit area) and is called as Pascal (Pa). A standard atmosphere, which is the atmospheric pressure at the standard sea level, is 1.01325×10^5 Pa. Pressure is also denoted in bars and 1 bar = 10^5 Pa.

For a homogeneous system at equilibrium, the pressure is uniformly the same throughout the system. For a system in mechanical equilibrium with its environment, the pressure is the same across the system's boundary.

1.5.2 TEMPERATURE

Temperature is an intensive variable that has its origin in thermodynamics. It is a measure of the physiological sensation of "hot" and "cold". The measurement of temperature is based on the fact that two systems brought into thermal contact will eventually reach the same state of "hotness", that is, a state of thermal equilibrium and will have the same value of temperature. This is the "zeroth" law of thermodynamics that can be stated as follows: If system A is in thermal equilibrium with system B (such as when brought in contact with each other) and system B is in thermal equilibrium with system C, then systems A and C are also in thermal equilibrium.

The zeroth law permits us to choose a test system called a thermometer to compare how "hot" the system of interest is and to determine its temperature.

The substances used in the thermometer should have a property that changes significantly with temperature and can be measured precisely. Most substances change their volume with temperature. Thus, the volume change can be calibrated to indicate the temperature change. A typical thermometric substance is a liquid (e.g., mercury, alcohol) contained in a small thin-walled glass bulb, which connects to a fine-bore capillary tube. The height of the liquid column in the capillary tube can then be calibrated to provide a scale to read the temperature. The property should also change linearly with temperature for easy measurements.

The change of the electrical resistance with temperature or the voltage from a thermocouple can also serve as a thermometer. The different measured parameters of the various thermometers provide the different temperature scales. Practical considerations require a thermometer to be sufficiently small so that it produces negligible effect on the system whose temperature is measured.

In a gas thermometer, a small volume of gas containing n_0 moles is enclosed in a bulb and either the volume change at constant pressure or pressure change at fixed volume can be used to measure the temperature changes. A constant volume gas thermometer is preferable since the pressure change with temperature can be accurately measured using a manometer. Use of gas at low pressures appears to provide a thermometric substance independent of the type of gas used.

1.5.2.1 Empirical Temperature θ

If X is the value of the thermometric property that changes with temperature parameter, for example, height of the mercury column in the capillary tube or pressure in a constant volume gas thermometer, then the ratio of the thermometric property X can be used to define the ratio of the temperature "θ". The temperature obtained in this manner is an empirical temperature, and its value depends on the particular thermometer used. When the thermometer is brought into thermal contact with heat sources A and B and if the thermometer reads X_A and X_B, respectively, we say that the empirical temperatures θ_A and θ_B of A and B are in the ratio

$$\frac{\theta_A}{\theta_B} = \frac{X_A}{X_B} \tag{1.2}$$

To obtain the empirical temperature scale, we need to assign a numerical value to some chosen heat source, for example, temperature of steam at atmospheric pressure. It is agreed upon that the triple point of water (equilibrium between ice, water and steam) be used as a standard heat source and assigned a particular value θ_{tp}. Thus, we write Eq. 1.2 as

$$\theta = \theta_{tp} \frac{X}{X_{tp}} \tag{1.3}$$

where θ is the temperature of the system to be measured, and X is the value of the thermometric substance when the thermometer is in thermal equilibrium with the system. θ_{tp} and the corresponding X_{tp} refer to the triple-point temperature and the value of X when the thermometer is at thermal equilibrium with a system of ice, water and steam at the triple point. From Eq. 1.2, we see that the ratio of the thermometric substance differs for different thermometers and the empirical temperature θ measured varies for different thermometers used.

1.5.2.2 Absolute Temperature T

For a gas thermometer, where we use the pressure of a constant volume gas at low pressure to indicate the temperature, we write Eq. 1.3 as

$$\theta = \theta_{tp}\left(\frac{p}{p_{tp}}\right) \tag{1.4}$$

It was found experimentally that $\dfrac{p}{p_{tp}}$ is independent of the type of gas used in the limit the amount of gas in the bulb (number of moles n_0 of it) approaching zero, that is, $\lim\limits_{n_0 \to 0}\left(\dfrac{p}{p_{tp}}\right)$ does not depend on the properties of the thermometric fluid. The temperature so obtained is known as the absolute temperature and is given as

$$T = T_{tp}\lim_{n_0 \to 0}\left(\frac{p}{p_{tp}}\right) \tag{1.5}$$

where T denotes the absolute temperature. It is measured in Kelvin (K).

1.5.2.3 Temperature in K and °C

Historically prior to the choice of the triple-point temperatures, the reference of ice water and steam water at 1 atmosphere pressure was used to determine an empirical temperature scale known as the Celsius scale. The temperature difference between ice and steam was chosen to be 100, that is, $T_s - T_i = 100$. Here, T_s and T_i denote the temperature of the reference steam water and ice water mixture. Thus, for a gas thermometer,

$$\frac{T_s}{T_i} = \frac{100 + T_i}{T_i} = \lim_{n_i \to 0}\frac{p_s}{p_i} \tag{1.6}$$

and solving for T_i gives

$$T_i = \frac{100}{\dfrac{p_s}{p_i} - 1}$$

Accurate measurement of $\dfrac{p_s}{p_i}$ gives its value to be 1.3661. Thus, $T_i = 273.15$ K. The temperature of the freezing point of water, which in degree Celsius is 0°C, is 273.15 in the Kelvin scale. Hence from Eq. 1.6, we write

$$T = 273.15\lim_{n_0 \to 0}\left(\frac{p}{p_i}\right)(\text{K})$$

We therefore shift the scale by 273.15 to convert °C to K, that is, $T(\text{K}) = \theta°(\text{C}) + 273.15$.

TABLE 1.1

Fixed Points for the International Practical Temperature Scale (1968)

Equilibrium Scale	T (K)	θ (°C)
Triple point of equilibrium hydrogen	13.81	−259.34
Boiling point of hydrogen at 25.76 atm.	17.042	−256.106
Boiling point of hydrogen at 1 atm.	20.28	−252.87
Boiling point of neon at 1 atm.	27.402	−246.048
Triple point of oxygen	54.361	−218.789
Boiling point of oxygen at 1 atm.	90.188	−182.962
Triple point of water	273.16	0.01
Boiling point of water at 1 atm.	373.15	100
Freezing point of zinc	692.73	419.58
Freezing point of silver	1,235.08	961.93
Freezing point of gold	1,337.58	1,064.43

Triple-point temperatures for different substances are given in Table 1.1 and can be used instead of the triple-point temperature of water.

The International Committee of Weights and Measures agreed in October 1968 on a number of fixed points in degree C and K as given in Table 1.1. Further interpolation procedures are also specified for obtaining the intermediate temperatures using a platinum resistance thermometer and platinum rhodium thermocouples.

1.6   STATE OF A SYSTEM: STATE VARIABLES/ THERMODYNAMIC PROPERTIES

We have so far considered the simple single-phase homogeneous system. For this simplest case, the state of the system is specified by the amount and type of matter, its volume or mass or moles, pressure and temperature. These are the state variables or thermodynamic properties of the system. The pressure and temperature are specified when the system is in mechanical equilibrium and thermal equilibrium, respectively. There may be more complex systems where there are different constituents or components in it with multiple phases (e.g., liquid, solid, vapor) being present. Additional state variables would be required to define the equilibrium state of the substance as the system becomes more complex.

1.7   CHANGE OF STATE OF A SYSTEM: QUASI-STATIC, REVERSIBLE AND CYCLIC PROCESSES

1.7.1   QUASI-STATIC PROCESS

There are various ways or paths by which the equilibrium state of a system could be changed. A process refers to a particular path causing the change. The system may not be in equilibrium at the different instants during the change; hence, the inter-mediate non-equilibrium states cannot be defined and the path or process cannot be specified. An important property of energy transfer process in thermodynamics is that it has to be quasi-static.

A quasi-static process is one in which the change of state is effected very slowly so that the state of the system, as well as the environment in which the system interacts with, is arbitrarily close to equilibrium at all times during the process. A process therefore goes through a series of equilibrium states. It may be noted that the equilibrium state corresponds to that of an isolated system when a definite invariant state is reached.

Real processes are not quasi-static since changes occur at finite rates. However, if the time scale of the change is long compared to the relaxation time of the system to equilibrate when perturbed, then the real processes can be approximated as being quasi-static. The state of the system and the environment must also be arbitrarily close to each other since finite gradients in the thermodynamic state will result in finite acceleration and non-uniformities in the system and the environment.

Equilibrium thermodynamics does not involve time. When time appears, it is to be understood that the rate is infinitesimally slow for the process to be quasi-static.

1.7.2 REVERSIBLE PROCESS

Reversible processes are of importance in thermodynamics. A reversible process is one in which both the system and the environment with which it interacts with are returned to their original states when the direction of the process is reversed. The system follows the same sequence of equilibrium states in the reverse direction.

A reversible process must necessarily be quasi-static, but a quasi-static process may not be reversible, for example, when a dissipative process like friction is present.

It should be noted that a system can always be made to return to its initial state, but a reversible path also requires that the environment is also returned to its original condition. Internal irreversibility is associated with irreversible processes that occur within the system, for example, diffusion of mass and heat when the system approaches overall equilibrium. External irreversibility is associated with the interaction between the system and the environment, for example, heat exchange across a finite temperature difference between the system and the environment. Chemical reactions among the various chemical species within the system also give rise to internal irreversibility. Irreversibility also results when dissipation processes are involved.

Reversible processes seem to be highly restrictive, but they are very important to provide a reference to assess real processes. Again the heat transfer process can be quasi-static when carried out infinitesimally slowly, and further for heat transfer to be reversible, the temperature difference across the boundary must be vanishingly small. Heat transfer across a finite temperature difference is irreversible and in fact constitutes the second law of thermodynamics.

1.7.3 CYCLIC PROCESS: EFFICIENCY AND COEFFICIENT OF PERFORMANCE

A sequence of processes in which the initial and final states are the same is called a thermodynamic cycle or a cyclic process. A cyclic process is necessary to run a heat engine or a refrigerator or heat pump for continuously generating work or pumping out or pumping in heat to a system.

We can determine the work done in the cyclic process by summing the work done in different processes constituting the thermodynamic cycle. The efficiency of an engine operating in a cycle is defined as ratio of the work done by it to the energy supplied. The performance of a refrigerator or heat pump is not defined by an efficiency but rather by the term Coefficient of Performance (COP), which is the ratio of the intended heat abstraction (for a refrigerator) or heat supply (for a heat pump) to the work required to effect the heat transfer.

The processes in a reversible engine can be reversed to make it operate as a refrigerator or a heat pump.

2 Equation of State

2.1 INTRODUCTION

It is found that not all the state variables are independent of each other. As an example, for a volume V m³ of nitrogen gas having a mass m kg at a pressure p, the temperature is fixed and cannot be varied arbitrarily. The relationship among the state variables $f(p, V, T, m$ or $n) = 0$ is called the equation of state of the substance. Given a substance (e.g., a gas), the equation of state is determined experimentally by measuring the p, V and T values for a given amount of m kg or n moles of it and empirically fitting an equation for the dependence among p, V, T and m or n.

We have seen that the variable pressure for a system can be defined only when the system is in mechanical equilibrium while the temperature is defined for a system when in thermal equilibrium. Thermodynamics deals with systems in equilibrium and the state variables define the equilibrium state of a system. The equation of state therefore describes the relationships of the variables of the system when the system is in thermodynamic equilibrium, that is, when it is in mechanical and thermal equilibrium.

2.2 EQUATION OF STATE FOR AN IDEAL GAS

The simplest substance is the so-called ideal gas. A real gas at sufficiently low pressure and high temperature behaves as an ideal gas. Low pressure and high temperature are relative to the critical pressure and critical temperature of the gas. At the low pressures, number density of molecules in the volume is such that the intermolecular potential energy is negligible as compared to the kinetic energies of the individual molecules.

The values of the critical pressure and temperature of some common gases are given in Table 2.1.

TABLE 2.1
Critical Pressure and Temperature

S. No.	Gas	Critical Pressure (Atmosphere)	Critical Temperature (K)
1	Nitrogen	33.5	126.0
2	Oxygen	49.7	154.3
3	Hydrogen	12.8	33.2
4	Carbon Monoxide	35	134
5	Helium	2.26	5.3
6	Argon	48	151

DOI: 10.1201/9781003224044-2

As can be observed, for normal temperatures and pressures, the ideal gas assumption should be valid.

The equation of state can be deduced for an ideal gas from the experimental observations of Boyle (1662), Gay-Lussac (1802) and Charles (1687). For a given amount of ideal gas, Boyle observed that at a given temperature, the product of pressure and volume is a constant, that is,

$$pV = A(m,T) \qquad (2.1)$$

where A depends on the amount of gas m and the temperature T. In 1802, Gay-Lussac showed that the ratio of the volumes of a given mass of gas at the temperatures of saturated steam and ice at a pressure of 1 atmosphere is a constant for all the gases studied, that is,

$$\frac{V_s}{V_i} = \text{Constant} \qquad (2.2)$$

where V_s and V_i denote the volume at temperatures of saturated steam and ice, respectively, at 1 atmosphere pressure. The constant was found to be 1.375, but later more precise measurements gave the value to be 1.36609. Gay-Lussac mentioned that his observations were also obtained by Charles in 1687. Gay-Lussac's observation given by Eq. 2.2 can be generalized to read

$$V = B(m, p)\, T \qquad (2.3)$$

where B depends on the amount of gas m and the pressure p. Combining Boyle's and the so-called Gay-Lussac or Charles' law, that is, Eqs. 2.1 and 2.3, gives

$$p\, B(m,p) = A(m,T) \,/\, T \qquad (2.4)$$

Since the left hand side in the above equation is a function of p and m, while the right hand side is a function of m and T, we must have both the sides equal to a function of m only, that is, $C(m)$. Thus,

$$p\, B(m, p) = A(m, T)/T = C(m)$$

and this on substituting the value of $A(m, T)$ from Eq. 2.1 as pV gives:

$$pV = C(m)T$$

The volume increases with m by the same factor when T and p are kept constant. So the value of $C(m)$ must be linearly dependent on m. Thus, we write

$$pV = mRT \qquad (2.5)$$

where R is a constant for a given gas and is known as the specific gas constant. Here, T is the absolute temperature in Kelvin as measured for the low-pressure gas by an ideal gas thermometer. Equation 2.5 gives a relationship between p, V, m and T, and is therefore an equation of state for an ideal gas. We can write Eq. 2.5 as

$$pv = RT \qquad (2.6)$$

where $v = V/m$ is the specific volume per unit mass. Further, since the density of the gas $\rho = 1/v$, we can also write Eq. 2.6 as

$$p = \rho RT \qquad (2.7)$$

If M is the molecular mass of the ideal gas in g/mole, then Eq. 2.5 can be written as

$$pV = n\, R_0 T \qquad (2.8)$$

where n is the number of moles given by $n = m\, /\, M$ and $R_0 = M \times R$. R_0 is referred to as the universal gas constant and is the same for all gases, its value being

$$R_0 = 8.314 \text{ J/(mole K)} \qquad (2.9)$$

Defining the specific volume on a molar basis as the volume per unit mole and denoting it by $\tilde{v}$ then $\tilde{v} = V\, /\, n$ and Eq. 2.8 gives:

$$p\tilde{v} = R_0 T \qquad (2.10)$$

If instead of mass or moles of a gas, we were to consider the number of molecules N in the ideal gas, Eq. 2.5 is written as

$$pV = N\, k\, T \qquad (2.11)$$

where k is a constant with dimensions of energy for a molecule per unit temperature and is given by

$$k = 1.381 \times 10^{-23} \text{ J/K} \qquad (2.12)$$

when the pressure p is given in Pascal and volume V is in m^3. k is referred to as the Boltzmann constant.

2.3   EQUATIONS OF STATE FOR REAL GASES

2.3.1   VIRIAL EQUATION OF STATE

The equation of state for an ideal gas is valid for vanishingly low pressures. Hence for higher pressures, correction terms are required. From experimental measurements of the pressure p and volume V at constant temperature, over a wide range of pressures, it is found that the product of pressure and specific volume per mole can be expressed in a power series in p or $\dfrac{1}{\tilde{v}}$, that is,

$$p\tilde{v} = a + bp + cp^2 + dp^3 + \cdots \tag{2.13}$$

$$p\tilde{v} = a\left(1 + \frac{b'}{\tilde{v}} + \frac{c'}{\tilde{v}^2} + \frac{d'}{\tilde{v}^3} + \cdots\right) \tag{2.14}$$

where $a, b, c, d, \ldots b', c', d'$, etc. are referred to as the virial coefficients and depend on temperature and the nature of the gas. For a gas at low pressure, that is, large specific volume, Eqs. 2.13 and 2.14 reduce to the ideal gas equation of state in the limit and the constant "a" must equal to R_0T.

The form of Eqs. 2.13 and 2.14 is called the virial equation of state. The virial coefficients will have the appropriate units depending on the units of p and $\tilde{v}$. For relatively low-pressure gases, it is sufficient to take the first couple of terms on the right side of Eqs. 2.13 and 2.14, for example, we write Eq. 2.14 as

$$p\tilde{v} = a\left(1 + \frac{b'}{\tilde{v}}\right) \tag{2.15}$$

giving $p\tilde{v}$ to be linearly dependent on $\frac{1}{\tilde{v}}$. The behavior of $p\tilde{v}$ versus pressure in the range of $0 < p < 40$ atm. at a temperature of 273.16 K is illustrated in Figure 2.1. It can be seen that the value of the product $p\tilde{v}$ for the different gases all asymptote to a value of $22.414 \dfrac{\text{L atm}}{\text{mole}}$ as the pressure tends to zero.

From the definition of temperature in Kelvin as measured in an ideal gas thermometer and given by Eq. 1.5 in Chapter 1

$$T = 273.16 \lim_{p, p_{tp} \to 0}\left(\frac{p}{p_{tp}}\right)$$

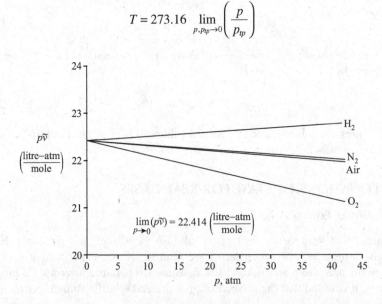

FIGURE 2.1   Variation of product $p\tilde{v}$ as a function of pressure for a few real gases.

where p_{tp} is the pressure measured in the ideal gas thermometer at the triple point of water. We can write the above equation as

$$T = 273.16 \lim_{p,p_{tp} \to 0} \left(\frac{p\tilde{v}}{p_{tp}\tilde{v}} \right)$$

$$\text{or} \lim_{p \to 0} \left(p\tilde{v} \right) = \left[\frac{\lim_{p_{tp} \to 0} \left(p_{tp}\tilde{v} \right)}{273.16} \right] T \tag{2.16}$$

However, $p_{tp} \approx 0.006$ atmosphere and $\lim_{p_{tp} \to 0} \left(p_{tp}\tilde{v} \right)$ from Figure 2.1 equals 22.414 L atm./mole at the triple point temperature of 273.16 K. We therefore obtain

$$p\tilde{v} = \left[\frac{22.414}{273.16} \right] T = 0.0820544 \, T$$

when the pressure $p \to 0$. Here, $R_0 = 0.0820544$ in liter atm./(mole K) and in terms of joules reduces to $R_0 = 8.314$ J/(mole K). Thus to the lowest order in pressure, we obtain the equation of state for an ideal gas.

Similarly, for higher pressures, we get

$$p\tilde{v} = R_0 T + bp$$

where the virial coefficient b is a function of temperature. As an example, for nitrogen, it is given by

$$b = 39.5 - \frac{1.00 \times 10^4}{T} - \frac{1.084 \times 10^6}{T^2} \tag{2.17}$$

where T is in Kelvin.

2.3.2 VAN DER WAAL'S EQUATION OF STATE

van der Waal in 1873 developed an equation of state for real gases in an attempt to correct the equation of state for an ideal gas. At higher pressures, the volume occupied by the gas molecules is no longer negligible. Thus, the molar volume in the ideal gas law is replaced by $(\tilde{v} - b)$. To account for intermolecular attraction, the pressure is written as $p + \frac{a}{\tilde{v}^2}$. The correction term $\frac{a}{\tilde{v}^2}$ for the attractive intermolecular force is based on the fact that it depends on the number of molecules (i.e., the molar density $\rho = \frac{1}{\tilde{v}}$) and also the intermolecular distance (which also depends on density $\rho = \frac{1}{\tilde{v}}$). Thus, the pressure correction term varies as $\frac{1}{\tilde{v}^2}$ The van der Waal equation of state is thus written as

$$p + \frac{a}{\tilde{v}^2} = \frac{R_0 T}{\tilde{v} - b} \tag{2.18}$$

with a and b to be determined experimentally. The constants a and b, however, actually depend on temperature and the values for a and b have to be determined for the particular regions of pressure and temperature of interest.

But since the critical constant temperature line (isotherm) on a $p - \tilde{v}$ diagram has zero slope and curvature at the critical point, we have the conditions

$$\left.\frac{\partial p}{\partial \tilde{v}}\right)_{T_c} = 0, \quad \left.\frac{\partial^2 p}{\partial \tilde{v}^2}\right)_{T_c} = 0 \tag{2.19}$$

where T_C denotes the critical temperature, from which the van der Waal coefficients a and b can be evaluated. Using Eqs. 2.19 and 2.18, we obtain

$$\left.\frac{\partial p}{\partial \tilde{v}}\right)_{T_c} = -\frac{R_0 T_c}{\left(\tilde{v}_C - b\right)^2} + \frac{2a}{\tilde{v}_C^3} = 0 \tag{2.20}$$

$$\left.\frac{\partial^2 p}{\partial \tilde{v}^2}\right)_{T_c} = \frac{2 R_0 T_c}{\left(\tilde{v}_C - b\right)^3} - \frac{6a}{\tilde{v}_C^4} = 0 \tag{2.21}$$

where the subscript "c" refers to the critical state. Solving for a and b yields

$$a = \frac{27}{64} \frac{R_0^2 T_C^2}{p_C}, b = \frac{R_0 T_C}{8 p_C}, \frac{p_C \tilde{v}_c}{R_0 T_C} = \frac{3}{8} \tag{2.22}$$

Thus from the critical point data for a given gas, the van der Waal coefficients a and b can be obtained.

Note that Eq. 2.22 gives $\dfrac{p_C \tilde{v}_C}{R_0 T_C} = \dfrac{3}{8} = 0.375$, whereas experimentally it is found that $\dfrac{p_C \tilde{v}_C}{R_0 T_C}$ has values in the range of 0.2–0.3. It would be more accurate to fit the experimental data in the region of p and T of interest to determine the coefficients a and b rather than using Eq. 2.22. The coefficients a and b for a few gases are given in Table 2.2.

The van der Waal equation of state is of historical interest since it represents the first attempt to correct the equation of state for an ideal gas taking into account the real gas effects.

2.3.3 BERTHELOT AND DIETERICI EQUATIONS OF STATE

There are other two-parameter equations of state where the two constants can similarly be obtained in terms of the critical pressure p_C and temperature T_C. Typical examples are the Berthelot and Dieterici equations, that is,

$$\text{Berthelot}: p = \frac{R_0 T}{\tilde{v} - b} - \frac{a}{T \tilde{v}^2} \tag{2.23}$$

$$\text{Dieterici}: p = \frac{R_0 T}{\tilde{v} - b} \exp\left(-\frac{a}{R_0 T \tilde{v}}\right) \tag{2.24}$$

TABLE 2.2
Constants for the van der Waal Equation of State a in bar/(m³/kmole)², b in m³/kmol

Substance	a	b	Substance	a	b
Acetylene	4.410	0.0510	Ethylene	4.563	0.0574
Air	1.358	0.0364	Helium	0.034	0.0234
Ammonia	4.233	0.0373	Hydrogen	0.0247	0.0265
Benzene	18.63	0.1196	Methane	2.285	0.0427
n-Butane	13.8	0.1196	Nitrogen	1.361	0.0385
Carbon dioxide	3.643	0.0427	Oxygen	1.369	0.0315
Carbon monoxide	1.463	0.0394	Propane	9.315	0.0900
Refrigerant 12 (CC2F3)	10.78	0.0998	Sulfur dioxide	6.837	0.0304
Ethane	5.575	0.0650	Water	5.507	0.0304

The constants a and b in the above equations can be obtained using the critical point conditions given by Eq. 2.19, to be

$$a = \frac{27R_0^2 T_C^2}{64p_C}, b = \frac{R_0 T_C}{8p_C} \tag{2.25}$$

for the Berthelot equation and

$$a = \frac{4R_0^2 T_C^2}{p_C e^2}, \ b = \frac{R_0 T_C}{p_C e^2} \tag{2.26}$$

where $e = 2.718$ for the Dieterici equation.

The Berthelot equation corrects for the attractive term in the van der Waal equation when the temperature is high and the kinetic energies of the molecules are large compared to the attractive potential energy. The correction term $\frac{a}{T\tilde{v}^2}$ would thus diminish with increase of temperature. The Dieterici equation was developed to give better agreement with the quantity $\frac{p_C \tilde{v}_C}{R_0 T_C}$ in Eq. 2.22 which is in considerable error in the van der Waal equation when compared with experiments.

2.3.4   REDLICH–KWONG EQUATION OF STATE

Another important equation of state used for high-pressure gases is the Redlich–Kwong equation given as

$$p = \frac{R_0 T}{\tilde{v} - b} - \frac{a}{T^{1/2}\tilde{v}(\tilde{v} + b)} \tag{2.27}$$

where the constants are determined as a function of the critical temperature and pressure as

TABLE 2.3
Constants a and b for the Redlich–Kwong Equation of State
a Is in bar $(m^3/kmole)^2(K)^{0.6}$, b is in $m^3/kmole$

Substance	a	b	Substance	a	b
Carbon dioxide	64.64	0.02969	Oxygen	17.38	0.02199
Carbon monoxide	17.26	0.02743	Propane	183.070	0.06269
Methane	32.19	0.02969	Refrigerant R12	214.03	0.02110
Nitrogen	15.59	0.02681	Water	142.64	0.02110

$$a = \frac{0.4275 R_0^{\,2} T_C^{\,2.5}}{p_C}, \quad b = \frac{0.0867 R_0 T_c}{p_C} \tag{2.28}$$

The Redlich–Kwong gives good results for the high-pressure region. The constants a and b for a few substances are given in Table 2.3.

2.4   COMPRESSIBILITY FACTOR AND GENERALIZED COMPRESSIBILITY CHART

The equations of state for the real gases, in general, are quite difficult to use. For most of the engineering problems, the compressibility chart provides a simple means to account for the dense gas effects. For an ideal gas $\dfrac{p\tilde{v}}{R_0 T} = 1$ and for a dense gas, we write

$$\frac{p\tilde{v}}{R_0 T} = Z \tag{2.29}$$

where Z is no longer equal to unity. Z is called the compressibility factor and varies with the state variables, that is, $Z(p, T)$. Different gases at the same p and T give different values of Z. However, if p and T are normalized with respect to the values at the critical state, that is, reduced temperature $T_R = \dfrac{T}{T_C}$ and reduced pressure $p_R = \dfrac{p}{p_C}$, then $Z(T_R, p_R)$ is a universal function valid for all gases. This is known as the theorem of corresponding states. The theorem states that "Any pure gas at the same reduced pressure and temperature should have the same compressibility factor".

Z is plotted against p_R for various constant values of T_R and is shown in Figure 2.2. The figure replaces the equation of state and we can find the p, v and T data from it for any gas.

The general procedure is to first calculate for the given p and T, the reduced pressure and temperature by normalizing with the critical pressure and temperature. The value of Z is obtained from the compressibility chart (Figure 2.2) and thereafter the value of the reduced molar volume $\tilde{v}_R$ is calculated. If, however, $\tilde{v}_R$ is given, then a value of $\tilde{v}_{R'} = \tilde{v}_R Z$, known as a pseudo reduced molar volume is used and plotted

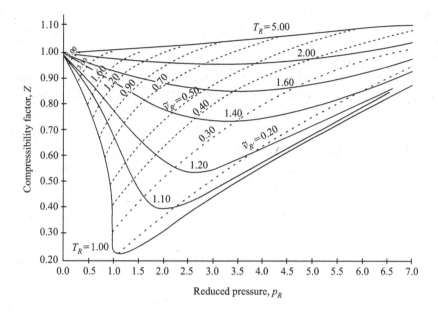

FIGURE 2.2 Generalized compressibility chart.

in the generalized chart instead of the reduced volume $\tilde{v}_R$ as it is more convenient. With $\tilde{v}_{R'}$ and p_R (or T_R), we can find the other missing state variable directly from the generalized compressibility charts in which constant $\tilde{v}_{R'}$ lines are plotted.

2.5 MIXTURE OF IDEAL GASES

For a mixture of several constituents or components of gases in a volume V at a given temperature T, the pressure that each constituent of gas would exert, if it were alone contained in the volume V, is its partial pressure. Let the partial pressure of the ith constituent in the mixture of gases be denoted by p_i, while the pressure of the mixture of gases is p. When each of the constituent gases in the mixture and the mixture of gases are ideal gases, then from the equation of state for an ideal gas $pV = nR_0T$, we get

$$\sum_{i=1}^{N} p_i = p \tag{2.30}$$

since the sum of the number of moles of each of the constituent gas is the total number of moles in the mixture. Equation 2.30 is Dalton's law of partial pressures for an ideal gas mixture.

Similarly we have Amagat's law of partial volume for a mixture of ideal gas. The law states that the volume of a mixture of ideal gas is equal to the sum of the partial volumes that each gas in the mixture would occupy if it existed in the mixture at the same temperature and pressure, that is,

$$\sum_{i=1}^{N} V_i = V \tag{2.31}$$

Dalton's law of partial pressure at constant volume and temperature and Amagat's law of partial volume at constant pressure and temperature are valid only when each of the constituent gases in the mixture and the mixture of gases are ideal gases.

3 First Law of Thermodynamics

3.1 STATEMENT OF THE FIRST LAW

The first law of thermodynamics states "the increase in the internal energy of a system is equaled to the heat transfer to the system minus the work done by the system". By convention, the heat transfer to and the work done by the system are considered to be positive. Thus

$$\Delta U = Q - W \tag{3.1}$$

Here, ΔU is the change in the internal energy, Q is the heat transfer and W is the work done. The first law is in essence of the law of conservation of energy.

Work is a familiar concept in mechanics, but "internal energy" and "heat" are novel to the first law.

3.2 INTERNAL ENERGY AND ADIABATIC WORK

Work done by a force is defined as the product of the force and the displacement in the direction of the force. In thermodynamics, we are mostly concerned with work associated with the volume changes of a system. If "p" denotes the pressure that the system exerts on its boundary and "dV" is the volume change, then "$p\,dV$" is the work done by the system when the system increases its volume by "dV".

The difference in the pressure across the boundary of the system must be infinitesimally small giving rise to a fully resisted motion of the boundary in order to define work done by the system. The rate of expansion is thus sufficiently slow to permit both the system and the environment that it interacts with to be in equilibrium at all times.

The work done by the system between two equilibrium states is given by the integral

$$W = \int_{V_1}^{V_2} p\,dV \tag{3.2}$$

The above integral is a path integral and the path $p(V)$ must be specified in order to evaluate the integral. Also between the same two states, the work done is different for different paths; we therefore say that the work is a path-dependent quantity.

DOI: 10.1201/9781003224044-3

There are different kinds of work other than "$p\,dV$" work and also the work may not necessarily be associated with the change in the configuration of the system (like the volume). For example, the work input to a system, such as by a paddle wheel as mechanical work, does not involve a change in the configuration of the system. This is generally defined as dissipative work where the mechanical work input to the system is converted via viscous dissipation to heat, hence increasing the internal energy of the system.

When the system is insulated and hence there is no heat exchange with the environment, the work done either by or on the system is referred to as adiabatic work. Experiments indicate that adiabatic work done by (or on) the system between two equilibrium states is the same for different adiabatic processes. From Eq. 3.1, we can write

$$\Delta U = -W_{ad} \tag{3.3}$$

where W_{ad} is the adiabatic work done by the system. Since W_{ad} is path-independent, the internal energy change between two states is also path-independent.

We can therefore define a state function "U" such that the change in the state function can be determined by measuring the adiabatic work input to the system.

Internal energy can be understood at a more fundamental level based on the atomistic view of matter. However, as per the first law, the internal energy difference between two equilibrium states can be defined via a macroscopic measurement of the adiabatic work input to the system.

3.3   HEAT

From experience, heat is transferred across a boundary where there is a temperature difference. But by measuring the work done and the change in internal energy, the heat transfer to the system can be obtained from the first law, that is,

$$Q = \Delta U + W = W - W_{ad} \tag{3.4}$$

Heat is thus a measure of the non-adiabatic nature of the system. Depending on the temperature gradient, the heat flow can be *to* or *from* the system. However, the direction of heat flow is always from hot to cold as required by the second law.

Since work is path-dependent while internal energy is not, the heat transfer is also path-dependent.

The first law in a differential form can be written following Eqs. 3.1 and 3.4 as

$$dU = \delta Q - \delta W$$

where dU denotes a perfect differential while δQ and δW are imperfect differentials being path-dependent. In applying the above differential form of first law to a cyclic process and integrating over a cycle, we get

$$\oint dU = \oint \delta Q - \oint \delta W$$

Since $\oint dU = 0$ as U is a path-independent state variable,

$$\oint \delta Q = \oint \delta W$$

This implies that in a cyclic process the sum of heat transfer is the same as the sum of work done or stated differently; the heat and work interactions are equal for a system undergoing a cyclic process. The equivalence between heat and work gives the mechanical equivalent of heat.

Historically, the amount of heat was measured by the calorie. A calorie is the heat required to raise the temperature of 1 g of pure water at 14.5°C by 1°C. The work equivalent of heat is 4.1858 J.

3.4 HEAT CAPACITY OF A SYSTEM

In the absence of phase changes, the heat transfer to a system results in a temperature rise. The heat capacity "C" of a system is defined as

$$C = \lim_{\Delta T \to 0} \frac{Q}{\Delta T} = \frac{\delta Q}{dT} \tag{3.5}$$

Since "Q" is a path-dependent quantity, δQ in Eq. 3.5 is not a total differential. Equation 3.5 should not be interpreted as the derivative of "Q" with respect to "T". Since δQ is path-dependent, we define the heat capacities for different heat transfer processes. For a constant volume process, we write the heat capacity at constant volume as

$$C_v = \lim_{\Delta T \to 0} \frac{\delta Q_v}{\Delta T} \tag{3.6}$$

where δQ_v is the heat transfer under constant volume conditions. Similarly, for a constant pressure process, we define heat capacity at constant pressure as

$$C_p = \lim_{\Delta T \to 0} \frac{\delta Q_p}{\Delta T} \tag{3.7}$$

3.4.1 HEAT CAPACITY AT CONSTANT VOLUME

For a simple hydrostatic system, the state can be specified by the variables p, V and T. If the equation of state is known, then it suffices to use any pair of the three variables, for example, T and V, p and T or p and V. Choosing T and V as independent variables, we write the internal energy as $U(T, V)$ and

$$dU(T,V) = \left(\frac{\partial U}{\partial T}\right)_V dT + \left(\frac{\partial U}{\partial V}\right)_T dV$$

and writing the first law in a differential form as

$$\delta Q = dU + \delta W = dU + p\, dV$$

we obtain

$$\delta Q = \left(\frac{\partial U}{\partial T}\right)_V dT + \left(\left(\frac{\partial U}{\partial V}\right)_r + p\right) dV \qquad (3.8)$$

In a constant volume process where $dV = 0$, the heat capacity is therefore

$$\frac{\delta Q_v}{dT} = \left(\frac{\partial U}{\partial T}\right)_V = C_v \qquad (3.9)$$

The internal energy can then be written as

$$U = \int C_V(T,V)dT$$

If C_V can be treated as constant for the temperature range of interest T_1 and T_2, then

$$U = C_v(T_2 - T_1) \qquad (3.10)$$

Using Eq. 3.9, we can also write Eq. 3.8 as

$$\delta Q = C_v dT + \left(p + \left(\frac{\partial U}{\partial V}\right)_T\right) dV \qquad (3.11)$$

3.4.2 HEAT CAPACITY AT CONSTANT PRESSURE

The first law could also be expressed in terms of an enthalpy function $H = U + pV$ instead of internal energy. The law can then be written as

$$\delta Q = dH - Vdp \qquad (3.12)$$

If we choose p and T as independent variables, that is, $H(p, T)$, we write

$$dH(p,T) = \left(\frac{\partial H}{\partial T}\right)_p dT + \left(\frac{\partial H}{\partial p}\right)_T dp \qquad (3.13)$$

Replacing dH in the first law (Eq. 3.12) by the expression in Eq. 3.13 gives

$$\delta Q = \left(\frac{\partial H}{\partial T}\right)_P dT - \left(V - \left(\frac{\partial H}{\partial p}\right)_T\right) dp \qquad (3.14)$$

Therefore for a constant pressure processes where $dp = 0$, the above gives

$$\delta Q_P = \left(\frac{\partial H}{\partial T}\right)_P dT$$

and in terms of heat capacity

$$C_p = \frac{\delta Q_p}{dT} = \left(\frac{\partial H}{\partial T}\right)_p \qquad (3.15)$$

If the heat capacity is specified, we can determine the enthalpy function as

$$H = \int C_p(T,p)dT \qquad (3.16)$$

For constant values of heat capacity in the temperature range of T_1 to T_2, we have the enthalpy function as

$$H = C_p(T_2 - T_1) \qquad (3.17)$$

3.4.3 RELATION BETWEEN HEAT CAPACITIES

If we write V as a function of p and T, that is, $V(p, T)$, we obtain

$$dV = \left(\frac{\partial V}{\partial T}\right)_p dT + \left(\frac{\partial V}{\partial p}\right)_T dp$$

Substituting the above in Eq. 3.8, we get

$$\delta Q = C_v dT + \left(p + \left(\frac{\partial U}{\partial V}\right)_T\right)\left(\left(\frac{\partial V}{\partial T}\right)_p dT + \left(\frac{\partial V}{\partial p}\right)_T dp\right)$$

For a constant pressure process where $dp = 0$, the above becomes

$$\frac{\delta Q_p}{dT} = C_p = C_v + \left(p + \left(\frac{\partial U}{\partial V}\right)_T\right)\left(\frac{\partial V}{\partial T}\right)_p$$

or

$$\frac{C_p - C_v}{\left(\dfrac{\partial V}{\partial T}\right)_p} = p + \left(\frac{\partial U}{\partial V}\right)_T \qquad (3.18)$$

which is the relation between the heat capacities at constant pressure and constant volume.

3.4.4 SPECIFIC HEATS

The heat capacity per unit mass (or mole) is the specific heat. Accordingly, the specific heats at constant volume and at constant pressure from Eqs. 3.9 and 3.15 are

$$c_v = \underset{\Delta T \to 0}{Lim}\left(\frac{du}{\Delta T}\right)_v = \left(\frac{du}{dT}\right)_v ; \tilde{c}_v = \left(\frac{d\tilde{u}}{dT}\right)_v$$

$$c_p = \underset{\Delta T \to 0}{Lim}\left(\frac{dh}{\Delta T}\right)_p = \left(\frac{dh}{dT}\right)_p ; \tilde{c}_p = \left(\frac{d\tilde{h}}{dT}\right)_p$$

Here, c_v and $\tilde{c}_v$ denote the specific heats at constant volume per unit mass and mole, respectively, while c_p and $\tilde{c}_p$ denote the specific heats at constant pressure per unit mass and mole. The specific internal energy per unit mass and per unit mole are denoted by $u = \dfrac{U}{m}$ and $\tilde{u} = \dfrac{U}{n}$, respectively, and similarly enthalpy per unit mass and per unit mole are denoted by $h = \dfrac{H}{m}$ and $\tilde{h} = \dfrac{H}{n}$.

3.5   INTERNAL ENERGY AND ENTHALPY FOR AN IDEAL GAS

For an ideal gas where $pV = nR_0T$,

$$\left(\frac{\partial V}{\partial T}\right)_p = \frac{nR_0}{p}$$

and assuming constant values of heat capacities C_p and C_V, we have using Eq. 3.10 and the definition of enthalpy function $H = U + pV$,

$$H_2 - H_1 = U_2 - U_1 + nR_0(T_2 - T_1)$$

and

$$C_p - C_v = n R_0 \tag{3.19}$$

$$\text{or } \tilde{c}_p - \tilde{c}_v = R_0; c_p - c_v = R$$

Substituting Eq. 3.19 in Eq. 3.18, we get

$$\left(\frac{\partial U}{\partial V}\right)_T = 0 \tag{3.20}$$

Thus for an ideal gas, the internal energy is a function of the temperature only. Similarly, it can be shown that the enthalpy function for an ideal gas is a function of its temperature alone.

3.6   EXPERIMENTAL VERIFICATION OF DEPENDENCE OF INTERNAL ENERGY ON TEMPERATURE, SPECIFIC VOLUME AND PRESSURE

Numerous attempts have been made to determine experimentally the dependence of internal of a gas on its specific volume. The early attempts were by Gay-Lussac and Joule at about the middle of the nineteenth century. Their experiments were based on the adiabatic-free expansion of a gas into vacuum where there is neither heat transfer nor work done by the gas. The first law indicates that the internal energy is a constant since there is neither heat nor work interaction. Careful measurements were made of the temperature change and within experimental accuracy, no temperature change was detected. Note that the experiment is extremely difficult to carry out because the temperature change, if any, is extremely small. More refined experiments were carried out subsequently with better experimental techniques to minimize heat losses and with higher accuracy of measurements for temperature. However, within the experimental error, no temperature change in the free expansion was observed. With no temperature change being observed after the free expansion, it may be concluded that the internal energy does not depend on the specific volume.

The experiments also attempted to measure the so-called Joule coefficient $\eta = \left(\dfrac{\partial T}{\partial v}\right)_u$, which should be zero if the specific internal energy u (internal energy per unit mass, viz. U/m) does not depend on the specific volume v (i.e., V/m). We can readily show that the Joule coefficient is related to $\left(\dfrac{\partial u}{\partial v}\right)_T$ from calculus. The cyclic rule gives

$$\left(\frac{\partial u}{\partial v}\right)_T \left(\frac{\partial v}{\partial T}\right)_u \left(\frac{\partial T}{\partial u}\right)_v = -1$$

Further, since $\left(\dfrac{\partial u}{\partial T}\right)_v = c_v$

where c_v is the specific heat capacity at constant volume, $c_v = \dfrac{dC_V}{dm}$ and using the reciprocal rule, we write

$$\eta = \left(\frac{\partial T}{\partial v}\right)_u = -\frac{1}{c_v}\left(\frac{\partial u}{\partial v}\right)_T \tag{3.21}$$

Since c_V is finite, if $\left(\dfrac{\partial T}{\partial v}\right)_u = 0$, then $\left(\dfrac{\partial u}{\partial v}\right)_T = 0$ and

$$u \neq f(v) \tag{3.22}$$

The early experiments were inconclusive. The idea of a precise measurement of the Joule coefficient was finally abandoned due to the difficulty of the direct measurement of temperature change of a gas in a free expansion.

The more modern method involves the measurement of a related quantity $\left(\dfrac{\partial u}{\partial p}\right)_T$ when a gas undergoes an isothermal expansion (where heat is transferred and work is done). The extensive experiments by Rossini and Frandsen in 1932 at the U.S. National Bureau of Standards indicated that the internal energy of a gas is a function of both temperature and pressure. They found no pressure and temperature range in which $\left(\dfrac{\partial u}{\partial p}\right)_T = 0$. They reported a value of $\left(\dfrac{\partial u}{\partial p}\right)_T = -6.08$ Joule/(mole atm.). In the pressure range of 1–40 atm., they observed $\left(\dfrac{\partial u}{\partial p}\right)_T$ is independent of pressure, that is,

$$\left(\frac{\partial u}{\partial p}\right)_T = f(T) \tag{3.23}$$

Thus in general $u = f(T) + g(T)\,p$

where $f(T)$ and $g(T)$ are functions of temperature.

3.7 EXPERIMENTAL VERIFICATION OF ENTHALPY TO BE INDEPENDENT OF PRESSURE FOR AN IDEAL GAS

To prove that the enthalpy is independent of pressure, Joule and Thomson carried out a related "throttling" experiment. In this experiment, steady flow of a gas in an insulated pipe is carried out with a flow restriction placed in the pipe to create a pressure drop. Temperatures upstream and downstream of the restriction were measured.

Application of the first law shows that the enthalpy is a constant in the throttling process. From the cyclic rule in calculus, we write

$$\left(\frac{\partial H}{\partial p}\right)_T \left(\frac{\partial p}{\partial T}\right)_H \left(\frac{\partial T}{\partial H}\right)_p = -1$$

Thus,

$$\left(\frac{\partial H}{\partial p}\right)_T = -\left(\frac{\partial H}{\partial T}\right)_p \left(\frac{\partial T}{\partial p}\right)_H = -C_p\,\mu \tag{3.24}$$

where

$$\mu = \left(\frac{\partial T}{\partial p}\right)_H$$

is called the Joule Thomson coefficient. The experiment measures the temperature across the restriction for given pressure difference and thus the Joule Thomson coefficient is determined. For an ideal gas, it was found that $\mu = 0$ and therefore the enthalpy is independent of pressure for an ideal gas and is just a function of its temperature. We will discuss further about the Joule Thomson coefficient in Chapter 8 on thermodynamic coefficients.

3.8 FIRST LAW APPLIED TO OPEN SYSTEMS

Open systems or control volumes permit mass as well as energy exchange across its boundary. Since in general, mass enters and leaves the open systems continuously, open systems are flow systems. Associated with the mass transfer, energy is also carried with the mass across the boundary. Further, displacement work is done on the system when mass is pushed in and work is done by the system in displacing the mass leaving the system.

An open system with fixed boundaries is shown in Figure 3.1. Mass and energy enter at Section 1 and leave at Section 2. Heat at a rate $\dot{Q}$ is supplied at the boundaries and work at the rate $\dot{W}_S$ is also delivered at the boundary as shown. Since the boundary does not move, the work is not displacement work as considered earlier; it could be shaft work from a turbine, electrical work, etc.

The increase of internal energy in the open system is related to the energy entering the control volume at Section 1 and energy leaving at Section 2, and the heat and work (Q and W_s) interactions at the boundary (control surface).

In order to derive a relationship between the rate of heat and work interactions at the boundary and the mass and energy entering and leaving the system, consider a flow into the open system at Section 1 of mass flow rate $\dot{m}_1$ at pressure p_1, specific volume v_1 and internal energy per unit mass u_1. Let the flow rate out of the control volume at Section 2 be $\dot{m}_2$ at pressure p_2, specific volume v_2 and specific internal energy u_2. Let the mass in the control volume at any instant of time be m and its specific internal energy (per unit mass) be u. If we were to consider a small mass Δm to flow through Section 1 into the control volume over a short duration Δt, the energy into the control volume over this short duration would be:

$$\Delta m(u_1 + p_1 v_1) = \dot{m}_1 \Delta t(u_1 + p_1 v_1) = \dot{m}_1 \, \Delta t \, h_1 \qquad (3.25)$$

Here, $p_1 v_1$ corresponds to the displacement work of unit mass of gas entering the open system at Section 1 at pressure p_1. The specific enthalpy per unit mass is $h = u + pv$.

Similarly, over the short duration Δt, energy outflow from the open system at Section 2 is

$$\dot{m}_2 \Delta t(u_2 + p_2 v_2) = \dot{m}_2 \, \Delta t \, h_2 \qquad (3.26)$$

Neglecting the flow velocities and consequently the kinetic energy within the open system, we have from first law the change in the internal energy of the open system as:

$$\Delta U = Q - W_s + \dot{m}_1 \Delta t h_1 - \dot{m}_2 \Delta t h_2 \qquad (3.27)$$

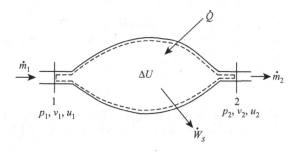

FIGURE 3.1 Open system.

Dividing the above equation by Δt gives:

$$\frac{dU}{dt} = \dot{Q} - \dot{W}_S + \dot{m}_1 h_1 - \dot{m}_2 h_2 \tag{3.28}$$

If in addition to the internal energy, the incoming and outgoing fluids at Sections 1 and 2 have kinetic energy from the flow velocities and potential energy from different heights above the datum, these kinetic and potential energies have to be included in Eq. 3.28. The kinetic energy within the system, however, is generally negligible compared to the enthalpies. But we may not be able to neglect the kinetic energy of the gas entering and leaving the system since the velocity at the entrance and exit could be large. Equation 3.28 therefore becomes

$$\frac{dU}{dt} = \dot{Q} - \dot{W}_S + \dot{m}_1 \left(h_1 + \frac{\bar{v}_1^2}{2} + gz_1 \right) - \dot{m}_2 \left(h_2 + \frac{\bar{v}_2^2}{2} + gz_2 \right) \tag{3.29}$$

Here, $\bar{v}_1$ and $\bar{v}_2$ denote the flow velocities at Sections 1 and 2 while z_1 and z_2 denote the heights above the datum. Equation 3.29 gives the rate of increase of internal energy in the control volume.

The rate of mass addition in the control volume is given by the conservation of mass:

$$\frac{dm}{dt} = \dot{m}_1 - \dot{m}_2 \tag{3.30}$$

For steady state, the rate of mass flowing in and out is the same, that is, $\dot{m}_1 = \dot{m}_2 = \dot{m}$ and $\frac{dU}{dt} = 0$. Substituting in Eq. 3.29 gives:

$$\frac{\dot{Q} - \dot{W}_S}{\dot{m}} = \left(h_2 + \frac{\bar{v}_2^2}{2} + gz_2 \right) - \left(h_1 + \frac{\bar{v}_1^2}{2} + gz_1 \right) \tag{3.31}$$

This is the steady flow energy equation for an open system.

If the datum or the potential energy of the incoming and outgoing streams is the same:

$$\frac{\dot{Q} - \dot{W}_S}{\dot{m}} = \left(h_2 + \frac{\bar{v}_2^2}{2} \right) - \left(h_1 + \frac{\bar{v}_1^2}{2} \right) \tag{3.32}$$

When the control volume is adiabatic and there is no work leaving the control surfaces (boundaries), that is, $\dot{Q} = 0$ and $\dot{W}_S = 0$, we get from the above equation:

$$h_1 - h_2 = \frac{\bar{v}_2^2}{2} - \frac{\bar{v}_1^2}{2} \tag{3.33}$$

which is the familiar form of the energy equation for adiabatic compressible flow. If the kinetic energies are negligible at the inlet and outlet, we have for steady flow

$$\dot{Q} - \dot{W}_S = \dot{m}(h_2 - h_1) \tag{3.34}$$

We find that we deal with the rates of the work and heat transfer in case of open systems. Representing w_S and q as work and heat interaction per unit mass, we can write Eq. 3.34 as

$$q - w_S = h_2 - h_1 \tag{3.35}$$

If we were to consider unit mass entering the system and follow its path through the system, we can write the first law for this unit mass following Eq. 3.12 as

$$dh = \delta q + v\,dp \tag{3.36}$$

Heat q refers to heat transfer per mass entering the open system. Integrating Eq. 3.36 between the entrance and exit of the open system, that is, between pressures p_1 and p_2, we write for a steady-state process

$$\Delta h = h_2 - h_1 = q + \int_{P_1}^{P_2} v\,dp \tag{3.37}$$

Thus,

$$-\int_{p_1}^{p_2} v\,dp = q - (h_2 - h_1) \tag{3.38}$$

and comparing the above with Eq. 3.35, we obtain

$$w_s = -\int_{P_1}^{P_2} v\,dp \tag{3.39}$$

To evaluate the above integral, the path of the change is to be specified.

4 Second Law of Thermodynamics

4.1 STATEMENTS OF THE SECOND LAW

The second law of thermodynamics is generally defined through two equivalent statements. The Clausius statement says that spontaneous heat flow from a cold to a hot reservoir is not possible. The Kelvin-Planck statement says that a perpetual motion machine of the second kind (PMM2) is not possible. By spontaneous, it is meant naturally without any energy input to aid the process. A reservoir is a body of large thermal capacity such that heat transfer from it will not change its temperature. A PMM2 is a machine that takes heat from a single heat reservoir and converts all the heat to mechanical work.

4.2 EQUIVALENCE OF KELVIN-PLANK AND CLAUSIUS STATEMENTS

To prove the equivalence of the two statements, we show that if one of the statements is not true, then the other is also not true, and vice versa. For example, if Clausius statement is not true, then spontaneous flow of heat from a cold reservoir to a hot reservoir is possible. We operate in a cyclic process a heat engine to take heat Q_1 from the hot reservoir at temperature T_1 and reject Q_2 to the cold reservoir at temperature T_2 as shown in Figure 4.1. We then let Q_2 to flow spontaneously back to the hot reservoir. The arrangement is a PMM2 that takes $Q_1 - Q_2$ from the hot reservoir and delivers work $W = Q_1 - Q_2$ (Figure 4.1). Thus, Kelvin-Planck statement is also not true.

If Kelvin-Planck statement is false, a PMM2 is possible. We let the PMM2 to take heat Q_1 from the hot reservoir and deliver work $W = Q_1$, as shown in Figure 4.2.

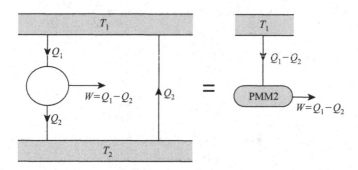

FIGURE 4.1 Violation of the Clausius statement equivalent to violation of the Kelvin-Plank statement.

DOI: 10.1201/9781003224044-4

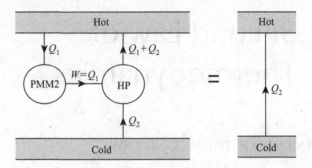

FIGURE 4.2 Violation of the Kelvin-Plank statement to a violation of the Clausius statement.

We use the work to operate a heat pump that extracts Q_2 from the cold reservoir and deliver $Q_{2+}W = Q_1 + Q_2$ to the hot reservoir. The combination of the PMM2 and the heat pump would result in the spontaneous heat flow of Q_2 from the cold to the hot reservoir (Figure 4.2). Thus, Clausius statement is not true. Hence, the statements are equivalent.

Note that the two statements of the second law should be considered as "highly improbable" rather than "not possible".

4.3 CARNOT'S PRINCIPLE

Using the second law, it can be shown that "no engine, operating in a cyclic process, can be more efficient than a reversible engine operating in a cyclic process between the same hot and cold reservoirs". This statement is referred to as Carnot's principle.

To prove Carnot's principle, consider an engine X and a reversible engine R both operating between a hot reservoir at temperature T_1 and a cold reservoir at T_2 as shown in Figure 4.3. It is to be noted that both the engines work in a cyclic process. Let engine X take heat Q_1 from the hot reservoir and reject $Q_{2'}$ to the cold reservoir (Figure 4.3). Similarly, operate R to take the same heat Q_1 from the hot reservoir but rejecting heat Q_2 to the cold reservoir. The work output of X and R would be $W_X = Q_1 - Q_{2'}$ and $W_R = Q_1 - Q_2$, respectively. The efficiency of X would be

$$\eta_X = \frac{W_X}{Q_1} = \frac{Q_1 - Q_{2'}}{Q_1} \text{ and similarly efficiency of } R \text{ is } \eta_R = \frac{W_R}{Q_1} = \frac{Q_1 - Q_2}{Q_1}$$

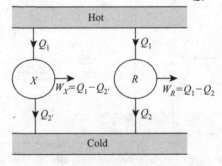

FIGURE 4.3 Operation of two engines between a set of reservoirs.

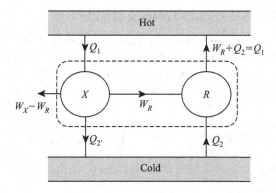

FIGURE 4.4 Irreversible engine X with efficiency greater than a reversible engine R.

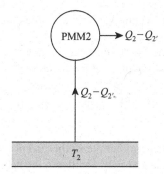

FIGURE 4.5 Irreversible engine with efficiency greater than reversible engine leads to PMM2.

If $\eta_x > \eta_R$, then $W_X > W_R$ and we can take W_R from the output of X to operate R backward as a reversible heat pump which takes Q_2 from the cold reservoir and delivers $W_R + Q_2 = (Q_1 - Q_2) + Q_2 = Q_1$ to the hot reservoir as shown in Figure 4.4. The combination of X and R would be a PMM2 that takes $Q_2 - Q_{2'}$ from the cold reservoir and delivers work $(W_X - W_R) = (Q_1 - Q_{2'}) - (Q_1 - Q_2) = Q_2 - Q_{2'}$ as shown in Figure 4.5. This violates the Kelvin-Planck statement and thus X cannot therefore be more efficient than R, that is, $\eta_x \le \eta_R$ where the equality sign applies if X is also a reversible engine.

4.3.1 Efficiencies of Reversible Engines

A corollary of Carnot's principle is that all reversible engines operating between the same hot and cold reservoirs must have the same efficiencies. The proof is straightforward. Denote R_1 and R_2 as the reversible engines. If $\eta_{R_1} > \eta_{R_2}$, then call R_1 engine X and we have just proven that $\eta_X = \eta_{R_1} \le \eta_{R_2}$. Alternatively if R_2 is more efficient than R_1, call R_2 engine X and we have proved that $\eta_X = \eta_{R_2} \le \eta_{R_1}$. Thus $\eta_{R_1} = \eta_{R_2}$ and similarly all reversible engines operating between the same hot and cold heat reservoirs will have the same efficiency, that is, $\eta_{R_1} = \eta_{R_2} = \eta_{R_3} = ----$.

4.4   HEAT TRANSFER AND TEMPERATURE

Perhaps the most fundamental cycle in thermodynamics is the reversible Carnot cycle. A cycle, it may be recalled, consists of a series of processes in which the system is returned to its initial state at the end of the cycle. The Carnot cycle consists of four reversible processes with two of them being isotherms and the other two adiabats as illustrated in Figure 4.6.

Heat Q_1 is absorbed in the isothermal process $a{\to}b$ wherein heat is taken in reversibly from the reservoir at temperature T_1, and heat Q_2 is rejected reversibly to the cold reservoir at temperature T_2 from $c{\to}d$. No heat is exchanged in the adiabatic processes from $b{\to}c$ and $d{\to}a$. The efficiency can be written as

$$\eta_C = \frac{Q_1 - Q_2}{Q_1} = 1 - \frac{Q_2}{Q_1} \tag{4.1}$$

Note that the Q's in the above equation have absolute values because the signs have already been taken into consideration.

If the working fluid in the Carnot cycle has n moles of an ideal gas, then for the isothermal process $a \to b$ taking place at temperature T_1:

$$Q_1 = W_{ab} = \int_{V_a}^{V_b} p\,dV = nR_0T_1\ln(V_b / V_a)$$

since the change in internal energy for the isothermal process in an ideal gas $\Delta U = 0$ and therefore from the first law $Q = W$. Similarly, the heat rejected Q_2 in the case of an ideal gas is

$$Q_2 = W_{cd} = \int_{V_c}^{V_d} p\,dV = n\,R_0T_2\ln(V_d / V_c)$$

Since $V_c > V_d$, the absolute value of Q_2 is written as

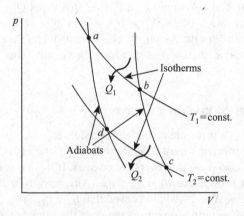

FIGURE 4.6   Carnot cycle.

$$Q_2 = n\,R_0 T_2 \ln\,(V_c\,/\,V_d)$$

The efficiency of the Carnot cycle is then

$$\eta_C = 1 - \frac{Q_2}{Q_1} = 1 - \frac{nR_0 T_2\,\mathrm{In}\,\dfrac{V_c}{V_d}}{nR_0 T_1\,\mathrm{In}\,\dfrac{V_b}{V_a}}$$

Further, the states "b" and "c", "d" and "a" are on the reversible adiabatic process. We can write the first law in a differential form for the adiabatic process as

$$\delta Q = 0 = dU + \delta W = dU + pdV$$

where pdV is the incremental work done in the slow quasi-static expansion or compression. The change in the internal energy dU for an ideal gas of mass m (corresponding to moles n) and specific heat c_v when the temperature changes by a small incremental value dT is $mc_v dT$ and therefore

$$mc_v dT = -pdV$$

For an ideal gas where $pV = mRT$ and $c_p - c_v = R$, R being the specific gas constant, that is,

$$c_v = \frac{R}{\gamma - 1}, \text{giving}$$

$$\frac{dT}{T} = -(\gamma - 1)\frac{dV}{V}$$

which on integrating gives $\ln T + (\gamma - 1)\ln V = C$, C being a constant.
    We therefore have

$$T_1 V_b{}^{\gamma-1} = T_2 V_c{}^{\gamma-1}\text{ and }T_2 V_d{}^{\gamma-1} = T_1 V_a{}^{\gamma-1}$$

Thus, $V_c/V_d = V_b/V_a$ and the efficiency becomes

$$\eta_C = 1 - \frac{Q_2}{Q_1} = 1 - \frac{T_2}{T_1} \tag{4.2}$$

The above equation gives the important result that the ratio of the heat transfer equals the ratio of the temperatures of the reservoirs for the Carnot cycle wherein an ideal gas is used as the working fluid. Since the equation of state for an ideal gas is used, the temperature T is the absolute temperature. Hence, the ratio of heat transfer in the Carnot cycle using an ideal gas is equaled to the ratio of the absolute temperatures of the reservoirs.

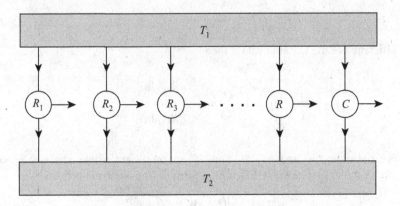

FIGURE 4.7   Set of reversible engines with one being a reversible Carnot engine.

From the corollary of Carnot's principle, we see that the efficiencies of all reversible engines operating between the same hot and cold reservoirs are the same, that is,

$$\eta_{R_1} = \eta_{R_2} = \eta_{R_3} = ---$$ (4.3)

If one of the reversible engines is a Carnot engine using an ideal gas as the working fluid, as shown by the last reversible engine in Figure 4.7, we obtain

$$\eta_{R_1} = \eta_{R_2} = \cdots = \eta_R = \eta_C = 1 - \frac{T_2}{T_1}$$ (4.4)

Thus, we obtain the general result that for a reversible engine, the ratio of the heat transfer must be equal to the ratio of the absolute temperatures of the reservoirs, that is, $\frac{Q_2}{Q_1} = \frac{T_2}{T_1}$ irrespective of what reversible cycle is used and what working fluid the reversible engine is based upon.

4.5   THERMODYNAMIC TEMPERATURE

In the last section, it was seen that the ratio of heat transfers from the reservoirs equals the ratio of the absolute temperatures of the reservoirs exchanging heat in reversible engines. This absolute temperature being proportional to heat transfer provides a thermodynamic definition of temperature, which is dealt with in the following. To be able to do so, we first prove that the efficiency of a reversible engine must depend on the temperature of <u>both</u> the hot and cold reservoirs it operates on. We specify the temperatures of the reservoirs to be some arbitrary temperature θ. A reversible engine R_1 operates between reservoirs at temperatures θ_1 and θ_2. Another reversible engine R_3 operates between reservoirs at temperatures θ_1 and θ_3 as shown in Figure 4.8.

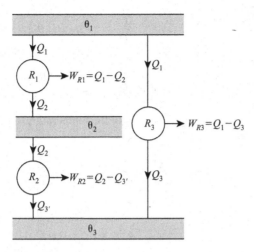

FIGURE 4.8 Reversible engines operating between reservoirs at temperatures θ_1, θ_2 and θ_3.

4.5.1 EFFICIENCY OF REVERSIBLE ENGINE DEPENDS ON TEMPERATURE OF BOTH RESERVOIRS

We operate both R_1 and R_3 to take the same heat Q_1 from the hot reservoir at temperature θ_1. The efficiencies of R_1 and R_3 would be $\eta_{R_1} = \dfrac{W_{R1}}{Q_1}, \eta_{R_3} = \dfrac{W_{R3}}{Q_1}$ where W_{R1} and W_{R3} are the work produced by R_1 and R_3. If the efficiency of a reversible engine depends only on the temperature of one reservoir, then $\eta_{R_1} = \eta_{R_3}$ and $W_{R1} = W_{R3}$ since both take Q_1 from reservoir at θ_1.

If the temperatures are such that $\theta_3 < \theta_2 < \theta_1$, then we can operate another reversible engine R_2 to take Q_2 from reservoir at θ_2 and deliver work W_{R2}, rejecting heat $Q_{3'}$ to reservoir θ_3. The combination of R_1 and R_2 is another reversible engine operating between reservoirs θ_1 and θ_3. The efficiency of $R_1 + R_2$ is then $\eta_{R_1 + R_2} = \dfrac{W_{R1} + W_{R2}}{Q_1}$ and will be greater than the efficiency of R_3 if the efficiency depends on a single reservoir at θ_1. This would violate the Carnot principle.

If $\eta_{R_1 + R_2} = \eta_{R_3}$, then $W_{R1} + W_{R2} = W_{R3}$ and $(Q_1 - Q_2) + (Q_2 - Q_{3'}) = Q_1 - Q_{3'} = Q_1 - Q_3$ or $Q_3 = Q_{3'}$. Thus, all reversible engines taking the same heat from the hot reservoir must also discharge the same amount of heat to the cold reservoir since their efficiencies are also the same in accordance with Carnot's principle. Thus, the efficiency of a reversible engine must depend on the temperature of both reservoirs.

4.5.2 THERMODYNAMIC TEMPERATURE RATIOS

With the efficiencies of reversible engines depending on both reservoirs, we can write the efficiencies of the three engines R_1, R_2 and R_3 as

$$\eta_{R_1}(\theta_1, \theta_2) = \frac{Q_1 - Q_2}{Q_1} = 1 - \frac{Q_2}{Q_1}; \ \frac{Q_2}{Q_1} = 1 - \eta_{R_1}(\theta_1, \theta_2) = f_{21}(\theta_2, \theta_1) \qquad (4.5)$$

$$\eta_{R2}(\theta_2,\theta_3)=\frac{Q_2-Q_3}{Q_2}=1-\frac{Q_3}{Q_2};\ \frac{Q_3}{Q_2}=1-\eta_{R2}(\theta_2,\theta_3)=f_{32}(\theta_3,\theta_2) \qquad (4.6)$$

$$\eta_{R3}(\theta_1,\theta_3)=\frac{Q_1-Q_3}{Q_1}=1-\frac{Q_3}{Q_1};\ \frac{Q_3}{Q_1}=1-\eta_{R3}(\theta_1,\theta_3)=f_{31}(\theta_3,\theta_1) \qquad (4.7)$$

where f_{21}, f_{32} and f_{31} are arbitrary functions of the temperatures of the reservoirs.
Since

$$\frac{Q_3}{Q_1}=\frac{Q_2}{Q_1}\cdot\frac{Q_3}{Q_2}$$

we see that

$$f_{31}(\theta_3,\theta_1)=f_{32}(\theta_3,\theta_2)\times f_{21}(\theta_2,\theta_1)$$

The left hand side of the above equation is a function of θ_3 and θ_1, whereas the right hand side is a function of θ_1, θ_2 and θ_3. This can be satisfied only if the function f_{31} (θ_3,θ_1) has the form

$$f_{31}(\theta_3,\theta_1)=\frac{\phi_3(\theta_3)}{\phi_1(\theta_1)}$$

where $\phi(\theta)$ is an arbitrary function of θ.
Hence,

$$f_{31}(\theta_3,\theta_1)=\frac{\phi_3(\theta_3)}{\phi_1(\theta_1)}=f_{32}(\theta_3,\theta_2)\times f_{21}(\theta_2,\theta_1)=\frac{\phi_3(\theta_3)}{\phi_2(\theta_2)}\times\frac{\phi_2(\theta_2)}{\phi_1(\theta_1)}=\frac{\phi_3(\theta_3)}{\phi_1(\theta_1)}$$

Without loss of generality, we can simply write the function $\phi(\theta)$ as θ itself. Thus, the ratio of heat transfer is a ratio of the temperature θ of the reservoirs, that is,

$$\frac{Q_2}{Q_1}=\frac{\theta_2}{\theta_1} \qquad (4.8)$$

The temperature θ is the absolute temperature since the heat rejected to the cold reservoir cannot be less than zero. We can therefore define the thermodynamic temperature on an absolute temperature scale based on heat transfer of a reversible heat engine.

4.5.3   THERMODYNAMIC OR ABSOLUTE TEMPERATURE SCALE

To obtain a temperature scale, one simply measures the efficiency of a reversible engine (hence Q_2/Q_1) between two chosen heat reservoirs. Different choices of a judicial system and various assignments of its temperature result in different temperature scales. As an example, for the Kelvin temperature scale, we keep the triple point of water as a standard reservoir and assign a temperature of 273.16 K for the triple point of water. By measuring the ratio of the heat transfer between a reservoir at temperature T to a reservoir at the triple point of water where we have assigned a temperature of 273.16 K, the temperature of the reservoir can be written as

$$T = 273.16 \left(\frac{Q}{Q_{t.p}} \right) \qquad (4.9)$$

We have replaced θ by T in the above for the absolute temperature. Q is the heat transfer from reservoir at temperature T while $Q_{t,p}$ is from the reservoir at the triple point temperature of 273.16 K. The advantage of the thermodynamic temperature scale is that it is independent of the type of reversible engine and the working fluids used in it.

4.6 CLAUSIUS INEQUALITY

Clausius inequality states that for a system undergoing a cycle and delivering work Ws while exchanging heat from various reservoirs at different temperatures, the sum of the heat transfer divided by the temperature of the reservoir over the cycle of operation is negative or zero, that is,

$$\sum_{\substack{\text{cycle;} \\ i=1,2...}} \frac{Q_i}{T_i} \leq 0 \qquad (4.10)$$

The cyclic system with the interacting reservoirs at temperatures T_1, T_2, ..., T_k is illustrated in Figure 4.9.

To prove Clausius inequality, we first arrange to have the system exchanging heat with only one heat reservoir at temperature T_0 using reversible engines and reversible heat pumps as indicated in Figure 4.10. Here, the total system within the boundary includes the original system S exchanging heat with reservoirs at temperatures T_1, T_2, ..., T_i, T_j and T_k and doing work W_S and the reservoirs at temperatures

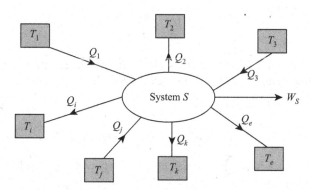

FIGURE 4.9 Schematic of cyclic system interacting with reservoirs and doing work W_S.

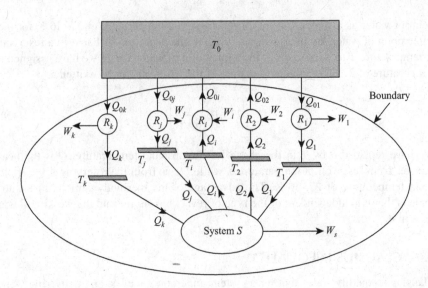

FIGURE 4.10 System exchanging heat with a single reservoir through reversible engines and reversible pumps.

$T_1, T_2, \ldots, T_i, T_j$ and T_k exchanging heat with only a single reservoir at temperature T_0 through reversible heat engines and heat pumps.

Applying the first law to the system within the boundary, we get the total work

$$W_T = \sum_{i=1,2,\text{--}} W_i + W_s = \sum_{\substack{\text{Cyclic:} \\ i=1,2,\text{--}}} Q_{oi} \qquad (4.11)$$

Here, $\sum W_i$ is the sum of the work done by reversible engines and on the heat pumps (all R_i's viz., $R_1, R_2, \ldots, R_i, R_j, R_k$) and W_S is the work done by the system. $\sum Q_{oi}$ is the net heat exchanges with the reservoir at T_0.

For the reversible engines or the reversible heat pumps, we have from the equality of ratios of heat transfer and temperatures

$$\frac{Q_{oi}}{Q_i} = \frac{T_0}{T_i}$$

where Q_i represents the heat transfers between the system and the reversible engine or pump at temperature T_i as shown in Figure 4.10. We get from the above equation

$$Q_{oi} = T_0 \frac{Q_i}{T_i} \qquad (4.12)$$

From second law, we have $\sum_{\substack{\text{cyclic} \\ i=1,2,\ldots}} Q_{oi} \leq 0$; otherwise we would have a PMM2 that violates the second law (Kelvin-Planck statement).

Thus $\displaystyle\sum_{\substack{\text{cyclic}\\ i=1,2,\ldots}} Q_{\text{o}i} = \sum_{\substack{\text{cyclic}\\ i=1,2,\ldots}} T_0 \frac{Q_i}{T_i} = T_0 \sum_{\substack{\text{cyclic}\\ i=1,2,\ldots}} \frac{Q_i}{T_i} \leq 0$

$$\text{or} \quad \sum_{\substack{\text{cyclic;}\\ i=1,2,\ldots}} \frac{Q_i}{T_i} \leq 0 \tag{4.13}$$

where the equality sign applies when the system undergoes a reversible cycle. If the system interacts with an infinite array of heat reservoirs whose temperatures differ by an infinitesimally small value dT with each supplying an infinitesimal heat δQ_i, we may write Clausius inequality as

$$\oint \frac{\delta Q}{T} \leq 0 \tag{4.14}$$

$$\text{or} \quad \oint \frac{\delta Q_{\text{rev}}}{T} = 0 \tag{4.15}$$

where δQ_{rev} denotes the reversible heat transfer.

We could also prove the equality sign for a reversible process by addressing the work done by a system. Since the system within the boundary in Figure 4.10 interacts with a single reservoir, the work cannot be positive; otherwise we would have a PMM2, which violates second law. However if the work is negative, that is, work done on the system, then it is possible to have such a system. The work, in this case, is dissipative work that heats up the system, which then transfers heat to the reservoir. Negative work means negative sum of heat transfer to the system and is acceptable. Now for a reversible cycle, the work cannot be dissipative work. So if the work cannot be positive (PMM2) and cannot be negative (reversible cycle), the work for a reversible cycle must be zero. Accordingly the sum of heat transfer must also be zero for a reversible cycle. We thus have the equality sign in the Clausius inequality for a reversible process.

4.7   ENTROPY

Since the integral of $\dfrac{\delta Q_{\text{rev}}}{T}$ over a cycle vanishes (Eq. 4.15), $\dfrac{\delta Q_{\text{rev}}}{T}$ is a total differential. Accordingly we define a state function

$$dS = \frac{\delta Q_{\text{rev}}}{T} \tag{4.16}$$

where S is called the entropy.

4.7.1   ENTROPY STATEMENT OF THE SECOND LAW

For an isolated system undergoing an irreversible process from states 1 to 2, we can seek a reversible path from states 2 to 1 and form a cycle. Thus,

$$\int_{1}^{2} \frac{\delta Q_{\text{irr}}}{T} + \int_{2}^{1} \frac{\delta Q_{\text{rev}}}{T} = \oint dS = \frac{\delta Q}{T} \le 0$$

and since $dS = \dfrac{\delta Q_{\text{rev}}}{T}$, the above gives

$$\Delta S = S_2 - S_1 \ge \int_{1}^{2} \frac{\delta Q_{\text{irr}}}{T} \tag{4.17}$$

For an isolated system where $\delta Q = 0$, the above yields

$$\Delta S)_{\text{isolated}} \ge 0 \tag{4.18}$$

with the equality sign applicable if the isolated system undergoes a reversible process. Equation 4.18 is another statement of the second law. It is equivalent to the Clausius and Kelvin-Planck statements. This can be easily demonstrated as follows.

4.7.2 EQUIVALENCE OF ENTROPY STATEMENT OF SECOND LAW AND CLAUSIUS AND KELVIN-PLANK STATEMENTS

Consider an isolated system consisting of a hot and a cold reservoir at temperatures T_1 and T_2 with $T_1 > T_2$ as shown in Figure 4.11. Let heat Q flow from the cold to the hot reservoir, that is, from T_2 to T_1. The entropy change of the isolated system (within the dotted boundary) is

$$\Delta S_{\text{isolated}} = \frac{Q}{T_1} - \frac{Q}{T_2} = \frac{Q}{T_2}\left(\frac{T_2}{T_1} - 1\right) \tag{4.19}$$

Since $T_1 > T_2$, $\Delta S_{\text{isolated}} < 0$ and thus heat flow from cold to hot reservoir is not possible as stated by Clausius.

To show the equivalence to the Kelvin-Planck statement, consider an isolated system consisting of a reservoir at temperature T and a PMM2 that takes heat Q from the

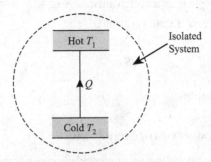

FIGURE 4.11   Isolated system comprising hot and cold reservoirs with $T_1 > T_2$.

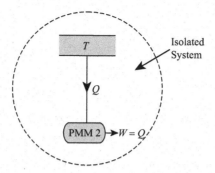

FIGURE 4.12 Isolated system consisting of reservoir and PMM2.

reservoir and delivers work $W = Q$. The reservoir and PMM2 constitute an isolated system (Figure 4.12).

The entropy change of the isolated system is

$$\Delta S = -\frac{Q}{T} < 0 \tag{4.20}$$

and thus violates the entropy statement of the second law as given by Eq. 4.18.

5 Entropy

5.1 ENTROPY BETWEEN TWO STATES

Entropy is an important state function. It was defined by Clausius inequality in Chapter 4 on the second law of thermodynamics as

$$dS = \left(\frac{\delta Q}{T}\right)_{rev}$$

To determine the entropy change between two equilibrium states, we need to integrate the above equation along a reversible path linking the two states. Since the entropy S is a state function, it is path-independent and it is not important what reversible path is chosen for its determination. It may be recalled that for a reversible path, all processes must be reversible, that is, both the mechanical and the heat transfer processes. The reversible processes must be quasi-static, that is, proceed through a series of equilibrium states.

5.2 PATH INDEPENDENCE

That entropy is independent of the path is shown by considering two states of a system of an ideal gas, namely, p_1, V_1, T_1 and p_2, V_2, T_2 with $T_1 = T_2 = T$. Thus, the two states lie on an isotherm $pV = nR_0T =$ constant and is shown in Figure 5.1. The environment is also considered to be at the same temperature ($T_0 = T$). The isothermal path is reversible and is chosen to determine the entropy change $\Delta S = S_2 - S_1$. Since the internal energy of an ideal gas depends only on the temperature, we have for the isothermal path $dU = 0$ and therefore the first law becomes $\delta Q_{rev} = pdV$. Thus,

$$\Delta S = \int \frac{\delta Q_{rev}}{T} = \int_{V_1}^{V_2} \frac{p}{T} dV = nR_0 \ln \frac{V_2}{V_1} \qquad (5.1)$$

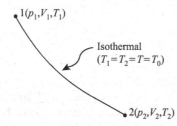

FIGURE 5.1   Isothermal path for an ideal gas at the same temperature as the environment.

DOI: 10.1201/9781003224044-5

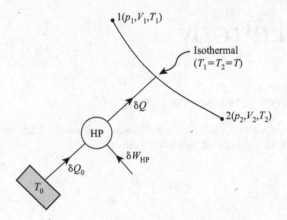

FIGURE 5.2 Isothermal path with environment at temperature T_0 different from path $T = T_1 = T_2$.

We have assumed that the system exchanges heat with the environment at temperature T_0 and $T_0 = T$. If $T_0 \neq T$, then a reversible heat pump is used to effect the reversible heat transfer as shown in Figure 5.2. Denoting δQ_0 as the reversible heat transfer during an infinitesimal part of the isothermal path from the environment at temperature T_0, we have from the second law

$$\frac{\delta Q_0}{T_0} = \frac{\delta Q}{T} = \frac{pdV}{T}$$

The entropy change in the environment is therefore

$$\Delta S_{\text{env}} = -\int \frac{\delta Q_0}{T_0} = -nR_0 \ln \frac{V_2}{V_1} \tag{5.2}$$

But

$$\Delta S_{\text{sys}} + \Delta S_{\text{env}} = 0$$

in accord with the second law. This gives the entropy change between states 1 and 2 as

$$\Delta S_{\text{sys}} = nR_0 \ln \frac{V_2}{V_1} \tag{5.3}$$

which is the same as Eq. 5.1 derived earlier.

We choose other paths between states 1 and 2: a reversible adiabatic path from 1 to 1' and a constant pressure heat addition path from 1' to 2 (Figure 5.3) or the reversible adiabatic path 1 to 1' followed by a constant volume heat addition path to give the same state 2 (Figure 5.4).

For the reversible adiabatic path 1 to 1', $\Delta S = 0$.

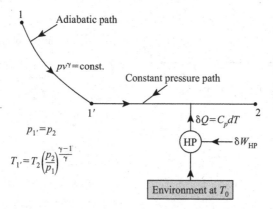

FIGURE 5.3 Path 1→2 comprising reversible adiabatic path 1→1' and constant pressure path 1'→2 with reversible heat pump HP to ensure the reversible path.

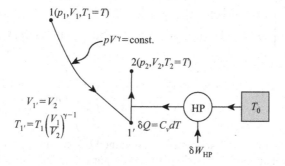

FIGURE 5.4 Reversible adiabatic path followed by a reversible constant volume path.

For the constant pressure path $1' \to 2$ in Figure 5.3,

$$\Delta S_{1' \to 2} = \int_{T_{1'}}^{T_2} \frac{C_p dT}{T} = C_p \ln \frac{T_2}{T_{1'}}$$

where $C_p dT$ is the reversible heat transfer from the environment during an infinitesimally small part of the constant path to make the path reversible. Hence,

$$\Delta S_{1' \to 2} = C_p \ln \left(\frac{p_1}{p_2} \right)^{\frac{\gamma-1}{\gamma}} = C_p \ln \left(\frac{V_2}{V_1} \right)^{\frac{\gamma-1}{\gamma}}$$

and similarly for the constant volume path $1' \to 2$ in Figure 5.4

$$\Delta S_{1' \to 2} = \int_{T_{1'}}^{T_2} \frac{C_v dT}{T} = C_v \ln \frac{T_2}{T_{1'}} = C_v \ln \left(\frac{V_2}{V_1} \right)^{\gamma-1}$$

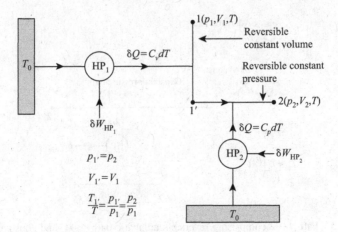

FIGURE 5.5 Path $1\rightarrow2$ comprising reversible constant volume process $1\rightarrow1'$ and reversible constant pressure process $1'\rightarrow2$ with incorporation of heat pumps.

Since $C_P\left(\dfrac{\gamma-1}{\gamma}\right)=nR_0$ and $C_V(\gamma-1)=nR_0$, we obtained the entropy change as

$$\Delta S = C_p \ln\left(\frac{p_1}{p_2}\right)^{\frac{\gamma-1}{\gamma}} = C_v \ln\left(\frac{V_2}{V_1}\right)^{\gamma-1} = nR_0 \ln\frac{V_2}{V_1} \tag{5.4}$$

in both the cases and is the same result as obtained before in Eqs. 5.1 and 5.3.

We can try yet another reversible path: a constant volume heat transfer from 1 to $1'$ followed by a constant pressure heat addition from $1'$ to 2 as shown in Figure 5.5.

The entropy change is written as

$$\Delta S = \Delta S_{1\rightarrow1'} + \Delta S_{1'\rightarrow2} = C_V \ln\frac{T_{1'}}{T} + C_P \ln\frac{T}{T_{1'}}$$

$$= (C_P - C_V)\ln\frac{T}{T_{1'}} = nR_0 \ln\frac{V_2}{V_1} \tag{5.5}$$

which is the same result obtained using different reversible paths from 1 to 2. Although we chose a particular system of ideal gas, the results can be generalized to any system.

5.3 GENERALIZED EXPRESSION FOR ENTROPY CHANGE

A convenient expression can be used to compute ΔS between two states. For a reversible path linking two states, the first law can be written as

$$dU = \delta Q_{rev} - pdV$$

where we write pdV for δW_{rev}. Since $\delta Q_{rev} = Tds$, we obtain

$$dS = \frac{1}{T}dU + \frac{pdV}{T}. \tag{5.6}$$

For a simple system described by variables p, V and T, we can choose any combination of two of the three variables as independent variables, that is, T, V; p, T; and p, V.

5.3.1 Entropy from Internal Energy Changes: (Variables T and V)

Choosing T, V as independent variables, we have $U(T, V)$ and $S(T, V)$ in Eq. 5.6 and we write

$$dU(T,V) = \left(\frac{\partial U}{\partial T}\right)_V dT + \left(\frac{\partial U}{\partial V}\right)_T dV \tag{5.7}$$

$$dS(T,V) = \left(\frac{\partial S}{\partial T}\right)_V dT + \left(\frac{\partial S}{\partial V}\right)_T dV \tag{5.8}$$

Substituting for $dU(T, V)$ from Eq. 5.7 in Eq. 5.6 and using Eq. 5.8,

$$dS(T,V) = \frac{1}{T}\left(\frac{\partial U}{\partial T}\right)_V dT + \frac{1}{T}\left(p + \left(\frac{\partial U}{\partial V}\right)_T\right)dV. \tag{5.9}$$

$\left(\frac{\partial U}{\partial T}\right)_T$ can be expressed in terms of p, V and T and once the equation of state is given, we can then determine it. Since dS is a total differential, the differentiability condition gives

$$\frac{\partial}{\partial V}\left(\left(\frac{\partial S}{\partial T}\right)_V\right)_T = \frac{\partial}{\partial T}\left(\left(\frac{\partial S}{\partial V}\right)_T\right)_V.$$

Thus,

$$\frac{\partial}{\partial V}\left(\frac{1}{T}\left(\frac{\partial U}{\partial T}\right)_V\right)_T = \frac{\partial}{\partial T}\left(\frac{1}{T}\left(p + \left(\frac{\partial U}{\partial V}\right)_T\right)\right)_V.$$

The above gives

$$\frac{1}{T}\left(p + \left(\frac{\partial U}{\partial V}\right)_T\right) = \left(\frac{\partial p}{\partial T}\right)_V. \tag{5.10}$$

and

$$\left(\frac{\partial U}{\partial V}\right)_T = T\left(\frac{\partial p}{\partial T}\right)_V - p. \tag{5.11}$$

$\left(\dfrac{\partial p}{\partial T}\right)_V$ can be determined if the equation of state $f(p,T,V)=0$ is specified. As

an example, for an ideal gas where $pV = nR_0T$, $\left(\dfrac{\partial p}{\partial T}\right)_V = \dfrac{nR_0}{V}$ and $\left(\dfrac{\partial U}{\partial V}\right)_T = 0$.
This implies that for an ideal gas, the internal energy is not a function of V and is a
function only of temperature, a result noted in Chapter 3.

The entropy from Eq. 5.9 using Eq. 5.10 becomes

$$dS(T,V) = \frac{1}{T}\left(\frac{\partial U}{\partial T}\right)_V dT + \left(\frac{\partial p}{\partial T}\right)_V dV \tag{5.12}$$

and for an ideal gas, the above reduces to

$$dS(T,V) = C_V \frac{dT}{T} + nR_0 \frac{dV}{V} \tag{5.13}$$

Hence, the entropy change from state 1 to state 2 for an ideal gas is

$$\Delta S(T,V) = \int_{T_1}^{T_2} C_V(T)\frac{dT}{T} + nR_0 \ln\frac{V_2}{V_1} \tag{5.14}$$

5.3.2   ENTROPY FROM ENTHALPY CHANGES: (VARIABLES P AND T)

If p and T are chosen as independent variables, it is more convenient to use
the enthalpy function $H = U + pV$ instead of internal energy U. In terms of
enthalpy, the first law is written as

$$dH = TdS + Vdp$$

and

$$dS(p,T) = \frac{1}{T}dH(p,T) - \frac{1}{T}Vdp \tag{5.15}$$

We can write $H(p,T)$ and $S(p,T)$ as

$$dH(p,T) = \left(\frac{\partial H}{\partial T}\right)_p dT + \left(\frac{\partial H}{\partial p}\right)_T dp \tag{5.16}$$

$$dS(p,T) = \left(\frac{\partial S}{\partial T}\right)_p dT + \left(\frac{\partial S}{\partial p}\right)_T dp \tag{5.17}$$

Combining the above, we obtain

$$dS(p,T) = \frac{1}{T}\left(\frac{\partial H}{\partial T}\right)_p dT - \frac{1}{T}\left(V - \left(\frac{\partial H}{\partial p}\right)\right)_T dp \tag{5.18}$$

The differentiability condition for dS gives

$$\frac{\partial}{\partial p}\left(\left(\frac{\partial S}{\partial T}\right)_p\right)_T = \frac{\partial}{\partial T}\left(\left(\frac{\partial S}{\partial p}\right)_T\right)_p$$

Thus,

$$\frac{\partial}{\partial p}\left(\frac{1}{T}\left(\frac{\partial H}{\partial T}\right)_p\right)_T = \frac{\partial}{\partial T}\left(-\frac{1}{T}\left(V-\left(\frac{\partial H}{\partial p}\right)_T\right)\right)_p$$

The above yields

$$\frac{1}{T}\left(V-\left(\frac{\partial H}{\partial p}\right)_T\right) = \left(\frac{\partial V}{\partial T}\right)_p . \tag{5.19}$$

Hence,

$$\left(\frac{\partial H}{\partial p}\right)_T = V - T\left(\frac{\partial V}{\partial T}\right)_p \tag{5.20}$$

$\left(\dfrac{\partial H}{\partial p}\right)_T$ can be evaluated if the equation of state is known. For an ideal gas, where, $pV = nR_0T$, $\left(\dfrac{\partial V}{\partial T}\right)_p = \dfrac{nR_0}{p}$ and $\left(\dfrac{\partial H}{\partial p}\right) = 0$, that is, the enthalpy is independent of pressure and is only a function of the temperature as reported in Chapter 3.

Using Eq. 5.20, Eq. 5.16 gives

$$dH(p,T) = C_p dT + \left(V - T\left(\frac{\partial V}{\partial T}\right)_p\right)dp \tag{5.21}$$

and for an ideal gas

$$dH(T) = C_p dT \tag{5.22}$$

The entropy can be obtained from Eqs. 5.18 and 5.20 as

$$dS(p,T) = C_p \frac{dT}{T} - \left(\frac{\partial V}{\partial T}\right)_p dp \tag{5.23}$$

For an ideal gas where $\left(\dfrac{\partial V}{\partial T}\right)_p = \dfrac{nR_0}{p}$, the above reduces to

$$dS(p,T) = C_p \frac{dT}{T} - nR_0 \frac{dp}{p}$$

or the change in entropy between two states 1 and 2 is

$$\Delta S(T,V) = \int_{T_1}^{T_2} C_p(T)\frac{dT}{T} - nR_0 \ln\frac{p_2}{p_1} \qquad (5.24)$$

5.3.3 ENTROPY CHANGES AS A FUNCTION OF HEAT CAPACITIES: (VARIABLES P AND V)

If p and V are chosen as independent variables, we can obtain $U(p,V)$ and $S(p,V)$ in a similar manner. We first write

$$dU(p,V) = \left(\frac{\partial U}{\partial p}\right)_V dp + \left(\frac{\partial U}{\partial V}\right)_p dV$$

$$dS(p,V) = \left(\frac{\partial S}{\partial p}\right)_V dp + \left(\frac{\partial S}{\partial V}\right)_p dV$$

Using the chain rule for four thermodynamic variables, namely, U, p, T and V, we have

$$\left(\frac{\partial U}{\partial T}\right)_V \left(\frac{\partial T}{\partial p}\right)_V \left(\frac{\partial p}{\partial U}\right)_V = 1 \qquad (5.25)$$

This chain rule for four variables is different from the cyclic rule for three variables, which for variables U, T and p is

$$\left(\frac{\partial U}{\partial T}\right)\left(\frac{\partial T}{\partial p}\right)\left(\frac{\partial p}{\partial U}\right) = -1$$

The derivation of the cyclic rule for the four variables can be seen as follows. For the four variables V, U, p and T, we can express U in terms of T and V and T in terms of V and p to give

$$dU = \left(\frac{\partial U}{\partial V}\right)_T dV + \left(\frac{\partial U}{\partial T}\right)_V dT. \qquad (5.26)$$

and

$$dT = \left(\frac{\partial T}{\partial V}\right)_p dV + \left(\frac{\partial T}{\partial p}\right)_V dp \qquad (5.27)$$

Substituting dT from Eq. 5.27 in Eq. 5.26, we get

$$dU = \left(\left(\frac{\partial U}{\partial V}\right)_T + \left(\frac{\partial U}{\partial T}\right)_V\left(\frac{\partial T}{\partial V}\right)_p\right)dV + \left(\frac{\partial U}{\partial T}\right)_V\left(\frac{\partial T}{\partial p}\right)_V dp \qquad (5.28)$$

However, U as a function of V and p gives

$$dU = \left(\frac{\partial U}{\partial V}\right)_p dV + \left(\frac{\partial U}{\partial p}\right)_V dp \qquad (5.29)$$

Equating the coefficients of dp in Eqs. 5.28 and 5.29

$$\left(\frac{\partial U}{\partial T}\right)_V \left(\frac{\partial T}{\partial p}\right)_V = \left(\frac{\partial U}{\partial p}\right)_V \qquad (5.30)$$

This gives the chain relation between four state variables as

$$\left(\frac{\partial U}{\partial T}\right)_V \left(\frac{\partial T}{\partial p}\right)_V \left(\frac{\partial p}{\partial U}\right)_V = 1$$

Using this relation, we can write

$$\left(\frac{\partial U}{\partial p}\right)_V = \left(\frac{\partial U}{\partial T}\right)_V \left(\frac{\partial T}{\partial P}\right)_V = C_v \left(\frac{\partial T}{\partial p}\right)_V \qquad (5.31)$$

We can also write from the relation $U = H - pV$

$$\left(\frac{\partial U}{\partial V}\right)_P = \left(\frac{\partial H}{\partial V}\right)_P - p = \left(\frac{\partial H}{\partial T}\right)_P \left(\frac{\partial T}{\partial V}\right)_P - p$$

$$= C_p \left(\frac{\partial T}{\partial V}\right)_P - p \qquad (5.32)$$

Thus, the internal energy change can be written as

$$dU(p,V) = C_v \left(\frac{\partial T}{\partial p}\right)_v dp + \left(C_p \left(\frac{\partial T}{\partial V}\right)_p - p\right) dV. \qquad (5.33)$$

The above equation can be integrated for the particular case of an ideal gas where $pV = nR_0T$ and $\left(\frac{\partial T}{\partial p}\right)_V = \frac{V}{nR_0}$ and $\left(\frac{\partial T}{\partial V}\right)_p = \frac{p}{nR_0}$ to give

$$dU = \frac{C_V}{nR_0}(Vdp + pdV) = -\frac{C_V}{nR_0}d(pV) = C_V dT \qquad (5.34)$$

Thus for an ideal gas, the internal energy is seen to be only a function of its temperature as obtained earlier.

Similarly, for the entropy, we write

$$\left(\frac{\partial S}{\partial p}\right)_V = \left(\frac{\partial S}{\partial T}\right)_V \left(\frac{\partial T}{\partial p}\right)_V = \frac{1}{T}\left(\frac{\partial U}{\partial T}\right)_V \left(\frac{\partial T}{\partial p}\right)_V$$

$$= \frac{C_V}{T}\left(\frac{\partial T}{\partial p}\right)_V \tag{5.35}$$

For the four variables S, V, T and p, we write

$$\left(\frac{\partial S}{\partial V}\right)_p = \left(\frac{\partial S}{\partial T}\right)_p \left(\frac{\partial T}{\partial V}\right)_p = \frac{1}{T}\left(\frac{\partial H}{\partial T}\right)_p \left(\frac{\partial T}{\partial V}\right)_p$$

$$= \frac{C_P}{T}\left(\frac{\partial T}{\partial V}\right)_p \tag{5.36}$$

Thus from Eqs. 5.35 and 5.36, we get

$$dS(p,V) = \frac{C_V}{T}\left(\frac{\partial T}{\partial p}\right)_V dp + \frac{C_P}{T}\left(\frac{\partial T}{\partial V}\right)_p dV \tag{5.37}$$

Knowledge of the heat capacities C_p and C_V and the equation of state will permit Eq. 5.37 to be integrated.

5.4   ENTROPY CHANGES FOR AN IDEAL GAS

If we consider an ideal gas where $\left(\frac{\partial T}{\partial p}\right)_V = \frac{V}{nR_0}; \left(\frac{\partial T}{\partial V}\right)_p = \frac{p}{nR_0}$ and substituting in Eq. 5.37 gives

$$dS(p,V) = C_V \frac{dp}{p} + C_p \frac{dV}{V}. \tag{5.38}$$

For constant heat capacities C_p and C_V, the above integrates to yield

$$pV^\gamma = Ce^{\frac{\Delta S}{C_V}} \tag{5.39}$$

where $\gamma = \dfrac{C_p}{C_V}$ is the ratio of the heat capacities and C is an integration constant.

For a constant entropy process (isentropic process) where $\Delta S = 0$; we obtain from Eq. 5.39, the isentropic relation for an ideal gas to be

$$pV^\gamma = \text{Constant} \tag{5.40}$$

6 Reversible Work, Availability and Irreversibility

6.1 REVERSIBLE WORK

It can be shown from Clausius inequality that the reversible work between two equilibrium states is the same for all reversible paths linking the two states. It also follows that the reversible work is the maximum work.

To prove the above, consider a real path X linking two equilibrium states. The system exchanges heat with the environment at a temperature T_0. For the real path X linking states 1 and 2, as shown in Figure 6.1, the first law can be written as

$$\Delta U = Q_X - W_X \tag{6.1}$$

where Q_X and W_X are the heat transfer to and work done by the system for path X. Similarly for a reversible path R linking the two states (Figure 6.1), we write

$$\Delta U = Q_R - W_R \tag{6.2}$$

We can form a cycle by going from state 1 to state 2 via a path X and return to state 1 by reversing along a reversible path R. For the cycle, we write

$$\Delta U = 0 = \left(Q_X - W_X\right) - \left(Q_R - W_R\right)$$

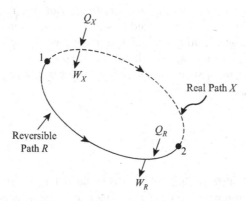

FIGURE 6.1 Reversible path R and irreversible path X linking states 1 and 2 with environment at temperature T_0.

DOI: 10.1201/9781003224044-6

Thus, $Q_X - Q_R = W_X - W_R$. From Clausius inequality

$$\sum_{\text{cycle}} \frac{Q}{T} \leq 0, \frac{Q_x}{T_0} - \frac{Q_R}{T_0} \leq 0$$

or

$$Q_X - Q_R \leq 0$$

The equality sign applies if X is also a reversible path. Hence, we obtain the result

$$Q_R \geq Q_X \tag{6.3}$$

that is, the heat transfer for the reversible path is greater than that for a real path. It follows that

$$W_X - W_R \leq 0$$

and

$$W_R \geq W_X \tag{6.4}$$

The equality sign in Eq. 6.4 applies if path X is also reversible. Further, the reversible work is the same for all reversible paths linking the two states. The work from a reversible path is therefore greater than that for a real (irreversible) path.

It also follows that the reversible work is the maximum work between two equilibrium states, that is, $W_{\text{rev}} = W_{\text{max}}$.

6.2 WORK FROM DIFFERENT REVERSIBLE PATHS BETWEEN TWO STATES

It would be of interest to show that the work done for different reversible paths between two equilibrium states is indeed the same. Consider for simplicity, an ideal gas whose state changes from p_1, V_1 and T_1 to p_2, V_2 and T_2. Let us take $T_1 = T_2 = T$; thus, states 1 and 2 lie on the same isotherm (Figure 6.2).

The work done by the system when its state changes from 1 to 2 along the isothermal path at temperature T is

$$W_{1\rightarrow2} = \int_{V_1}^{V_2} p\,dV = nR_0 T \ln \frac{V_2}{V_1} \tag{6.5}$$

If the temperature T of the isothermal process in the system is different from the temperature of the environment T_0 and if $T > T_0$, then a reversible heat pump is required to transfer heat reversibly to the system during the isothermal expansion as shown in Figure 6.2 to maintain reversibility. The heat pump work is given by

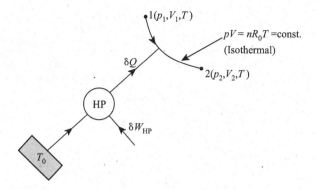

FIGURE 6.2 Isothermal path 1→2 of system with reversible heat pump to transfer heat reversibly from environment at temperature T_0.

$$W_{HP} = \int \frac{dQ}{\text{COP}} \qquad (6.6)$$

In the above expression, the coefficient of performance (COP) of the reversible heat pump is the ratio of heat supplied by the heat pump to the system at temperature T from the reservoir at T_0 to the work done on the heat pump. For reversible heat exchange Q with the system at temperature T and Q_0 from the reservoir at temperatures T_0, respectively, the coefficient of performance is

$$\text{COP}_{HP} = \frac{Q}{Q - Q_0} = \frac{T}{T - T_0} = \frac{1}{\left(1 - \dfrac{T_0}{T}\right)}$$

Substituting in Eq. 6.6 gives

$$W_{HP} = \int p \, dV \left(1 - \frac{T_0}{T}\right) = nR_0 T \ln \frac{V_2}{V_1} - nR_0 T_0 \ln \frac{V_2}{V_1}$$

The work from the heat pump is supplied to the system and is negative. Thus, the reversible work from $1 \to 2$ is

$$W_{1 \to 2} = nR_0 T \ln \frac{V_2}{V_1} - \left(nR_0 T \ln \frac{V_2}{V_1} - nR_0 T_0 \ln \frac{V_2}{V_1}\right)$$

$$= nR_0 T_0 \ln \frac{V_2}{V_1} \qquad (6.7)$$

If we have an alternate path for the change 1→2; a reversible constant volume from $1 \to 1'$ and a reversible constant pressure from $1' \to 2$ with the incorporation of heat pumps, as shown in Figure 6.3,

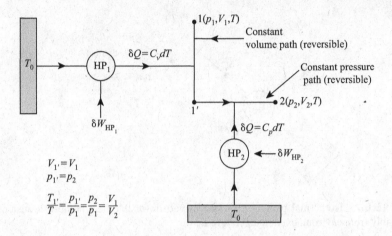

FIGURE 6.3 Path $1 \rightarrow 2$ comprising constant volume path $1 \rightarrow 1'$ and constant pressure path $1' \rightarrow 2$ with reversible heat pumps HP_1 and HP_2 to ensure the reversible path.

the work done for the first heat pump HP_1 to maintain a reversible constant volume path is

$$W_{HP_1} = \int \frac{C_V dT}{COP} = \int_T^{T_{1'}} C_v \, dT \left(1 - \frac{T_0}{T}\right) = C_V(T_{1'} - T) - C_v T_0 \ln\frac{T_{1'}}{T} \tag{6.8}$$

Here, C_V is the heat capacity of the ideal gas in the system at constant volume. From $1' \rightarrow 2$, the work of the heat pump HP_2 is

$$W_{HP_2} = \int_{T_{1'}}^T \frac{C_p dT}{COP} = \int_{T_{1'}}^T C_p \, dT \left(1 - \frac{T_0}{T}\right) = C_p(T - T_{1'}) - C_p T_0 \ln\frac{T}{T_{1'}} \tag{6.9}$$

with C_P being the heat capacity of the system at constant pressure.

The expansion work in the constant pressure process $1' \rightarrow 2$ is

$$p_2(V_2 - V_{1'}) = nR_0 T \left(1 - \frac{V_1}{V_2}\right) \tag{6.10}$$

Thus, the total work $1 \rightarrow 1' \rightarrow 2$ is

$$W_{1 \rightarrow 1' \rightarrow 2} = -\left(C_v(T_{1'} - T) - C_v T_0 \ln\frac{T_{1'}}{T}\right) - \left(C_p(T - T_{1'}) - C_p T_0 \ln\frac{T}{T_{1'}}\right) + nR_0 T\left(1 - \frac{V_1}{V_2}\right) \tag{6.11}$$

Note that the heat pump work is taken to be negative as the work is done *on* the reversible path $1 \rightarrow 1' \rightarrow 2$. Since for the constant pressure process $1' \rightarrow 2$ for the ideal gas $\dfrac{T_{1'}}{T} = \dfrac{V_1}{V_2}$, the above reduces to

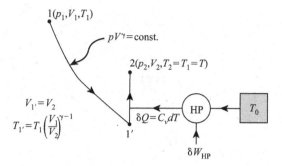

FIGURE 6.4 Reversible path 1→2 comprising of reversible adiabatic path 1→1′ and constant volume path 1′→2 with a reversible heat pump for the process to be reversible.

$$W_{1\to 1'\to 2} = nR_0T_0\ln\frac{V_2}{V_1} \tag{6.12}$$

which is the same result as obtained previously in Eq. 6.7.

Consider a third alternate reversible path from 1 to 2 with a reversible adiabatic (isentropic) path $1 \to 1'$ followed by a constant volume heat transfer from $1'$ to 2 where $V_{1'} = V_2$. This path is illustrated in Figure 6.4.

The work in the adiabatic expansion $1 \to 1'$ is $U_1 - U_{1'} = C_v(T - T_{1'})$. For the constant volume heat transfer path $1' \to 2$, the heat pump work is

$$W_{\text{HP}} = \int_{T_{1'}}^{T} C_v\, dT\left(1 - \frac{T_0}{T}\right) = C_V(T - T_{1'}) - C_V T_0\ln\frac{T}{T_{1'}} \tag{6.13}$$

The total work $1 \to 1' \to 2$ is therefore

$$W_{1\to 1'\to 2} = C_V T_0\ln\frac{T}{T_{1'}} = C_V T_0\left(\frac{V_2}{V_1}\right)^{\gamma-1} = nR_0T_0\ln\frac{V_2}{V_1} \tag{6.14}$$

This is the same as obtained for the other reversible paths. Thus, this simple example, using ideal gases, illustrates that the work for all reversible paths between two equilibrium states is the same. The choice of the particular case of an ideal gas does not influence the generality of the proof.

6.3 REVERSIBLE WORK OF A SYSTEM INTERACTING WITH ENVIRONMENT: AVAILABILITY Φ

An expression for the reversible or the maximum work can be obtained from the first law. If Q_{rev} is the heat transfer in the reversible path, we write

$$W_{\text{rev}} = W_{\text{max}} = Q_{\text{rev}} - \Delta U \tag{6.15}$$

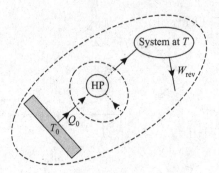

FIGURE 6.5 Change of state of a system reversibly with use of reversible heat pump.

Consider a system that takes Q_0 from the environment at temperature T_0 while going from state 1 to state 2. If the temperature of the system differs from T_0, then to transfer Q_0 to the system reversibly requires the use of a reversible engine (or reversible heat pump).

Taking the system and environment as an isolated system, as illustrated in Figure 6.5, the second law gives

$$\Delta S_{\text{isolated}} = \Delta S_{\text{system}} + \Delta S_{\text{env}} = \Delta S_{\text{system}} - \frac{Q_0}{T_0} \tag{6.16}$$

where we have put $\Delta S_{\text{env}} = -\dfrac{Q_0}{T_0}$

But for the reversible heat transfer in the isolated system

$$\Delta S_{\text{isolated}} = 0$$

Thus,

$$Q_0 = T_0 \Delta S_{\text{system}}$$

and from the first law applied to the system, we write

$$\Delta U = Q_0 - W_{\text{rev}}$$

or

$$W_{\text{rev}} = W_{\text{max}} = T_0 \Delta S_{\text{sys}} - \Delta U = -\Delta(U - T_0 S) \tag{6.17}$$

$$= -\Delta \Phi \tag{6.18}$$

where $\Phi = U - T_0 S_{\text{sys}}$ is defined as the availability function. Note that the availability function Φ is not a state function since it depends also on the temperature of the environment. Equation 6.18 states that the maximum work obtained between two equilibrium states of a system that interacts with the environment corresponds to the decrease of the availability function.

6.4 REVERSIBLE WORK OF A SYSTEM INTERACTING WITH RESERVOIR AND ENVIRONMENT

If the system takes Q_R from a heat reservoir at temperature T_R as well as Q_0 from the environment at T_0, then the second law gives

$$\Delta S_{\text{isolated}} = \Delta S_{\text{sys}} + \Delta S_{\text{res}} + \Delta S_{\text{env}} = \Delta S_{\text{sys}} - \frac{Q_R}{T_R} - \frac{Q_0}{T_0} = 0 \qquad (6.19)$$

where

$$\Delta S_{\text{res}} = -\frac{Q_R}{T_R}; \quad \Delta S_{\text{env}} = -\frac{Q_0}{T_0};$$

Solving for Q_0, we get

$$Q_0 = T_0 \Delta S_{\text{sys}} + T_0 \Delta S_{\text{res}} = T_0 \Delta S_{\text{sys}} - Q_R \frac{T_0}{T_R} \qquad (6.20)$$

and substituting the above in the first law gives

$$W_{\max} = (Q_0 + Q_R) - \Delta U = T_o \Delta S_{\text{sys}} + Q_R \left(1 - \frac{T_0}{T_R}\right) - \Delta U$$

$$= -\Delta(U - T_0 S_{\text{sys}}) + Q_R \left(1 - \frac{T_0}{T_R}\right) \qquad (6.21)$$

or

$$W_{\max} = -\Delta\Phi + Q_R \left(1 - \frac{T_0}{T_R}\right) \qquad (6.22)$$

Note that to transfer heat reversibly from the environment and reservoir to the system, reversible engines and heat pumps must be used. Further, if we transfer Q_R from the reservoir to the system, the heat received by the system is not Q_R. To transfer Q_R from the reservoir to the system, we first operate a reversible engine to take Q_R from the reservoir and discharge Q_R (T_0/T_R) to the environment. Then, we use a reversible heat pump to take the heat from the environment and deliver it to the system. In other words, we do not bypass the environment and go directly from the reservoir to the system via a reversible engine. Figure 6.6 illustrates the system interacting with the reservoir and the environment.

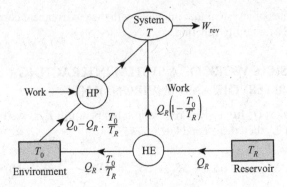

FIGURE 6.6   System interacting with reservoir and environment.

It is important to remember what is specified in the above. We specify two states of the system; hence ΔS_{Sys} and ΔU. We specify the heat transfer from the reservoir Q_R. However, we do not specify the heat taken from the environment Q_0. The amount of heat Q_0 varies with the magnitude of W_{max} and is defined in Eq. 6.20.

6.5   REVERSIBLE WORK WHEN SYSTEM CHANGES ITS VOLUME

If the system changes its volume between states 1 and 2, that is, $\Delta V = V_1 - V_2$, then displacement work $p_0 \Delta V$ is done by the system on the environment. This displacement work would not be available for other purposes. Thus if we are interested in the maximum useful work, then we have to subtract $p_0 \Delta V$ from W_{max}, that is,

$$W_{\substack{max \\ useful}} = W_{max} - p_0 \Delta V$$

We can include $p_0 \Delta V$ in the availability function Φ and define

$$\Phi^* = \left(U - T_0 S_{sys} + p_0 V_{sys} \right) \tag{6.23}$$

Thus, Eq. 6.22 is written as

$$W_{\substack{max \\ useful}} = -\Delta \Phi^* + Q_R \left(1 - \frac{T_0}{T_R} \right) \tag{6.24}$$

Similarly, we modify Eq. 6.18 if the displacement work is taken into consideration, that is,

$$W_{\substack{max \\ useful}} = -\Delta \Phi^* \tag{6.25}$$

6.6   IRREVERSIBILITY OF A SYSTEM UNDERGOING A PROCESS

If the work obtained in an actual process between states 1 and 2 is W_{actual}, we can compare this with the maximum work W_{max} given by Eqs. 6.18 and 6.22. We define the irreversibility I as

$$I = W_{\text{actual}} - W_{\text{max}} \qquad (6.26)$$

where I now serves as a qualitative measurement of how efficient a certain actual process is. As defined in Eq. 6.26, the irreversibility $I \leq 0$ with the equality sign holding good if the actual process is also reversible.

The irreversibility can also be expressed in terms of entropy change of the universe. Assuming both the actual and reversible processes take the same Q_R and Q_0 from a heat reservoir at T_R and the environment at T_0, respectively; $\Delta S_{\text{res}} = -\dfrac{Q_R}{T_R}$ and $\Delta S_{\text{env}} = -\dfrac{Q_0}{T_0}$ Note that the entropy change of the environment and the reservoir do not depend on the process that the system undergoes.

Writing the first law for the actual process in the system, we obtain

$$W_{\text{actual}} = -T_0 \Delta S_{\text{env}} - T_R \Delta S_{\text{res}} - \Delta U \qquad (6.27)$$

For the reversible process,

$$W_{\text{rev}} \text{ or } W_{\text{max}} = Q_0 + Q_R - \Delta U$$

However, for a reversible process between the same two states,

$$\Delta S_{\text{universe}} = \Delta S_{\text{sys}} + \Delta S_{\text{env}} + \Delta S_{\text{res}}$$

$$= -\frac{Q_0}{T_0} + \Delta S_{\text{sys}} + \Delta S_{\text{res}} = 0$$

which gives

$$Q_0 = T_0 \Delta S_{\text{sys}} + T_0 \Delta S_{\text{res}}$$

Thus,

$$W_{\text{max}} = T_0 \Delta S_{\text{sys}} + T_0 \Delta S_{\text{res}} - T_R \Delta S_{\text{res}} - \Delta U$$

The irreversibility then becomes

$$I = W_{\text{actual}} - W_{\text{max}}$$

$$= \left(-T_R \Delta S_{\text{res}} - T_0 \Delta S_{\text{env}} - \Delta U \right) - \left(T_0 \Delta S_{\text{sys}} + T_0 \Delta S_{\text{res}} - T_R \Delta S_{\text{res}} - \Delta U \right)$$

$$= -T_0 \left(\Delta S_{sys} + \Delta S_{res} + \Delta S_{env} \right)$$

$$= -T_0 \Delta S_{univ} \tag{6.28}$$

Since

$$\Delta S_{univ} \geq 0$$

the irreversibility

$$I \leq 0 \tag{6.29}$$

Entropy is a useful state function that permits one to assess quantitatively the loss in work potential of a real thermodynamic process.

6.7 TWO EXAMPLES ILLUSTRATING IRREVERSIBILITY

6.7.1 EXPANSION OF AN IDEAL GAS INTO VACUUM

Consider the example of the expansion of an ideal gas into vacuum inside an insulated chamber as shown in Figure 6.7. Let n moles of a perfect gas be at initial state (p_1, V_1) at temperature T_1 to the left of the partition, which separates the gas from the vacuum on the right side of volume $V_2 - V_1$. The partition is removed and the gas expands into the vacuum and the final state is (p_2, V_2) at temperature T_2.

From the first law, we have for the expansion $\Delta U = 0$ and hence $T_2 = T_1$ since no work is done in the expansion into a vacuum. For an ideal gas whose equation of state is $pV = nR_0 T$, we have $p_2 = \dfrac{p_1 V_1}{V_2}$. Thus, the final state is $p_2 = \dfrac{p_1 V_1}{V_2}$, $T_2 = T_1$. The entropy change of the system is $\Delta S_{sys} = nR_0 \ln \dfrac{V_2}{V_1}$. The maximum work for the expansion is obtained for an environmental temperature T_0 as

$$W_{max} = -\Delta \Phi = -\Delta U + T_0 \Delta S_{sys} = nR_0 T_0 \ln(V_2 / V_1) \tag{6.30}$$

Thus, irreversibility of the process is

$$I = W_{act} - W_{max} = 0 - nR_0 T_0 \ln \frac{V_2}{V_1} = -nR_0 T_0 \ln \frac{V_2}{V_1} \tag{6.31}$$

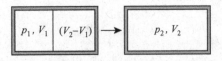

FIGURE 6.7   Expansion of an ideal gas into vacuum.

since the actual work W_{act} of the process of expansion in vacuum is zero.

We can alternatively find I from $I = -T_0 \Delta S_{universe}$. Since the system is completely isolated, it is the universe and thus

$$\Delta S_{universe} = \Delta S_{sys} = nR_0 \ln \frac{V_2}{V_1}$$

The irreversibility is therefore $I = W_{act} - W_{max} = -nR_0 T_0 \ln \frac{V_2}{V_1}$, which is the same given in Eq. 6.31.

It is instructive to obtain the above results from basic considerations. Given state 1 (p_1, V_1) at temperature T_1 and state 2 (p_2, V_2) at temperature T_2, where $T_2 = T_1$ and $p_2 = \frac{p_1 V_1}{V_2}$. Since states 1 and 2 lie on the same isotherm $T_1 = T_2 = T = $ constant, the total work of the reversible isothermal process is

$$\int_{\substack{V_1 \\ pV=C}}^{V_2} p\, dV = \int_{V_1}^{V_2} nR_0 T \frac{dV}{V} = nR_0 T \ln \frac{V_2}{V_1}$$

If the environment $T_0 < T$, then a reversible heat pump must be used for the heat transfer (Figure 6.8). The work of the heat pump is

$$W_{HP} = \int \left(\frac{p\,dV}{1/(1-T_0/T)} \right) = nR_0 T \ln \frac{V_2}{V_1} - nR_0 T_0 \ln \frac{V_2}{V_1}$$

The maximum work is for the reversible path and is

$$W_{max} = W - W_{HP} = nR_0 T \ln \frac{V_2}{V_1} - \left[nR_0 T \ln \frac{V_2}{V_1} - nR_0 T_0 \ln \frac{V_2}{V_1} \right]$$

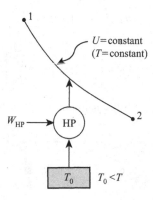

FIGURE 6.8 Expansion of an ideal gas into vacuum from state 1 to state 2.

$$= nR_0T_0\ln\frac{V_2}{V_1}$$

Thus, the irreversibility is

$$I = W_{act} - W_{max} = 0 - nR_0T_0\ln\frac{V_2}{V_1} = -nR_0T_0\ln\frac{V_2}{V_1}$$

This is the same result obtained from the availability function.

6.7.2 Cooling of a Cup of Hot Coffee

We can consider another example of the cooling of a cup of hot coffee at T_1 to room temperature T_0 illustrated in Figure 6.9.

The process is constant volume heat transfer from the coffee to the environment. The heat exchange takes place across a finite temperature and the process is irreversible. The actual work of irreversible cooling is zero, $W_{actual} = 0$. However, we cool the coffee reversibly and obtain some work.

From first law analysis, the heat transfer is

$$Q = \Delta U = mc(T_0 - T_1) \tag{6.32}$$

where m is the mass of coffee and c is the specific heat.

The entropy change of the coffee is

$$\Delta S_{coffee} = mc\ln\frac{T_0}{T_1} \tag{6.33}$$

The maximum work that can be obtained between the two states can be found from the availability function, that is,

$$W_{max} = -\Delta\Phi = -\Delta U + T_0\Delta S_{coffee}$$

$$= -mc(T_0 - T_1) + T_0mc\ln\frac{T_0}{T_1}$$

$$= -mcT_0\left[\left(\frac{T_1}{T_0} - 1\right) + \ln\frac{T_0}{T_1}\right]$$

Since $W_{actual} = 0$, the irreversibility is obtained as

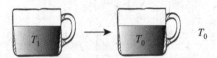

FIGURE 6.9   Cooling of a cup of hot coffee in the environment.

$$I = W_{\text{act}} - W_{\text{max}} = -mcT_0\left[\left(\frac{T_1}{T_0}-1\right)+\ln\frac{T_0}{T_1}\right] \tag{6.34}$$

Alternatively, we can find the irreversibility from entropy considerations, that is,

$$I = -T_0\Delta S_{\text{universe}} = -T_0\left(\Delta S_{\text{coffee}} + \Delta S_{\text{env}}\right)$$

with $\Delta S_{\text{coffee}} = mc\ln\frac{T_0}{T_1}$ and $\Delta S_{\text{env}} = \frac{mc(T_1 - T_0)}{T_0}$

Thus,

$$I = -mcT_0\left[\left(\frac{T_1}{T_0}-1\right)+\ln\frac{T_0}{T_1}\right] = mcT_0\left[\left(1-\frac{T_1}{T_0}\right)+\ln\frac{T_1}{T_0}\right]$$

which is the same result obtained earlier in Eq. 6.34.

We can also do the problem from basic considerations. We can find the maximum work by reversibly cooling the coffee at constant volume using a reversible engine between the hot coffee and the environment as shown in Figure 6.10.

The work of the reversible heat engine is

$$W_{\text{HE}} = \int_{T_1}^{T_0} -mc\,dT\left(1-\frac{T_0}{T}\right)$$

$$= -mcT_0\left(1-\frac{T_1}{T_0}\right)+mcT_0\ln\frac{T_0}{T_1}$$

$$= -mcT_0\left[\left(1-\frac{T_1}{T_0}\right)+\ln\frac{T_1}{T_0}\right]$$

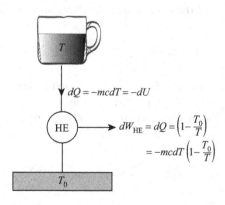

FIGURE 6.10 Reversible engine between system and environment.

Since constant volume heat transfer does not produce any work, the maximum work is just the work from the reversible heat engine. This gives

$$I = W_{\text{act}} - W_{\text{max}} = mcT_0\left[\left(1-\frac{T_1}{T_0}\right)+\ln\frac{T_1}{T_0}\right]$$

which is the same result as obtained earlier in Eq. 6.34.

6.8   IRREVERSIBILITY IN OPEN SYSTEMS

Consider rate of heat transfer $\dot{Q}_0$ to an open system from the environment at T_0 and $\dot{Q}_R$ from a reservoir at T_R. Thus, $\dot{Q} = \dot{Q}_0 + \dot{Q}_R$. For the process in the open system where $\dot{m}_1$ and $\dot{m}_2$ are the mass flow rates entering and leaving the open system with specific enthalpies h_1 and h_2, respectively, and $\dot{U}$ is the rate at which the internal energy of the system increases, the non-pdV work done from the first law for open systems (Chapter 3) gives

$$\dot{W}_{\text{actual}} = (\dot{Q}_0 + \dot{Q}_R)_{\text{act}} + \dot{m}_1 h_1 - \dot{m}_2 h_2 - \dot{U} \qquad (6.35)$$

The kinetic and potential energies of the gases at the entry and exit are not considered in the above. For the maximum rate of work done, all processes must be reversible. Thus to transfer heat reversibly from the environment and the reservoir to the open system, reversible heat pump and engine must be used. The work required by the heat pump and the work obtained from the engine are included in the work from the system. The first law is now written as

$$\dot{W}_{\text{max}} = \left(\dot{Q}_0 + \dot{Q}_R\right)_{\text{rev}} + \dot{m}h_1 - \dot{m}_2 h_2 - \dot{U}$$

For maximum work, all processes must be reversible, thus giving

$$\dot{S}_{\text{universe}} = \dot{S}_{\text{sys}} + \dot{S}_{\text{env}} + \dot{S}_{\text{res}} = 0$$

The net entropy change of the environment is written as

$$\dot{S}_{\text{env}} = -\frac{\dot{Q}_0}{T_0} - \dot{m}_1 s_1 + \dot{m}_2 s_2$$

where s_1 and s_2 are the specific entropies entering and leaving the system. The environment loses entropy when the mass carrying the entropy $\dot{m}_1 s_1$ enters the system.

Similarly, the environment gains entropy when the mass is discharged from the system to the environment. We therefore write

$$\dot{S}_{\text{universe}} = \dot{S}_{\text{sys}} - \frac{\dot{Q}_0}{T_0} - \dot{m}_1 s_1 + \dot{m}_2 s_2 - \frac{\dot{Q}_R}{T_R} = 0$$

where the rate of entropy change in the reservoir of temperature T_R with rate of heat transfer across it is $-\dfrac{\dot{Q}_R}{T_R} = \dot{S}_{res}$

Solving for $\dot{Q}_0$, we get

$$\dot{Q}_0 = T_0 \dot{S}_{sys} - \left[\dot{m}_1 s_1 - \dot{m}_2 s_2 - \frac{\dot{Q}_R}{T_R} \right] T_0$$

Substituting the above expression for $\dot{Q}_0$ into the first law, we write

$$\dot{W}_{max} = T_0 \dot{S}_{sys} + \dot{Q}_R \left(1 - \frac{T_0}{T_R} \right) + \dot{m}_1 (h_1 - T_0 s_1) - \dot{m}_2 (h_2 - T_0 s_2) - \dot{U} \qquad (6.36)$$

Defining the open system availability function Ψ per unit mass as

$$\psi = h - T_0 s \qquad (6.37)$$

similar to the closed system availability function as

$$\Phi = U - T_0 S_{sys} \qquad (6.38)$$

the maximum rate of work done by the open system is written as

$$\dot{W}_{max} = -\dot{\Phi} + \dot{Q}_R \left(1 - \frac{T_0}{T_R} \right) + \dot{m}_1 \Psi_1 + \dot{m}_2 \Psi_2 \qquad (6.39)$$

For steady flow where $\dot{m}_1 = \dot{m}_2 = \dot{m}$ and $\dot{U} = 0, \dot{S}_{sys} = 0$, we write

$$w_{max} = q_R \left(1 - \frac{T_0}{T_R} \right) - \Delta \Psi \qquad (6.40)$$

where $\Delta \Psi = \Psi_2 - \Psi_1$, w_{max} is the maximum work per unit mass and q_R is also the heat transfer per unit mass that flows through the system.

The irreversibility can be obtained from Eqs. 6.35 and 6.36, that is,

$$\dot{I} = \dot{W}_{actual} - \dot{W}_{max} \qquad (6.41)$$

or $\dot{I} = \left[\dot{Q}_0 + \dot{Q}_R + \dot{m}_1 h_1 - \dot{m}_2 h_2 - U \right] - \left[T_0 \dot{S}_{sys} + \dot{Q}_R \left(1 - \frac{T_0}{T_R} \right) + \dot{m}_1 (h_1 - T_0 s_1) - \dot{m}_2 (h_2 - T_0 s_2) - \dot{U} \right]$

Noting that $\dot{S}_{env} = -\dfrac{\dot{Q}_0}{T_0} - \dot{m}_1 s_1 + \dot{m}_2 s_2$ and $\dot{S}_{res} = -\dfrac{\dot{Q}_R}{T_R}$

The above equation reduces to

$$\dot{I} = -T_0\left(\dot{S}_{sys} + \dot{S}_{env} + \dot{S}_{res}\right) = -T_0\dot{S}_{universe} \tag{6.42}$$

The above result is similar to the irreversibility obtained for a closed system except that for open systems, it is the rate of irreversibility.

7 Thermodynamic State Functions

7.1 INTRODUCTION

Heat and work interactions among different systems depend on the relevant changes of the thermodynamic states and the thermodynamic properties of the systems. The thermodynamic properties could either be determined experimentally or from the fundamental properties of the molecules through statistical thermodynamics. The latter is due to the large number of molecules in a system behaving as a group in a statistical way. Some of the thermodynamic properties can be measured directly, for example, pressure, temperature, volume and mass. Specific heats can also be obtained via calorimetric measurements. However, thermodynamic state functions, which define the equilibrium state of a system, irrespective of how the equilibrium state is reached, are not directly measurable. Thermodynamic relationships must be developed to permit the state functions such as internal energy and entropy to be obtained from the measurable thermodynamic properties or variables. The thermodynamic properties that cannot be directly measured could also be expressed in terms of those that can be readily measured. In this chapter, the state functions and thermodynamic properties are discussed and the relationships between them are derived.

7.2 STATE FUNCTIONS

7.2.1 INTERNAL ENERGY

The internal energy U was introduced in connection with the first law, that is,

$$dU = \delta Q - \delta W \tag{7.1}$$

where δQ is the heat transfer to the system and δW is the work done by the system when the system undergoes a process. The change in internal energy dU is path-independent (i.e., a total differential), whereas δQ and δW are path-dependent quantities.

A simple hydrostatic system is characterized by its internal energy U and its mass m. For simplicity, we initially consider a single component system and its state is characterized by the internal energy per unit mass (specific internal energy) $u = \dfrac{U}{m}$.

7.2.2 ENTROPY

The thermodynamic state function entropy was defined from the second law as

$$dS = \frac{dQ_{\text{rev}}}{T} \tag{7.2}$$

DOI: 10.1201/9781003224044-7

dQ_{rev} in the above expression is the reversible heat transfer between two neighboring states and is path-independent since the reversible heat transfer is the same for all reversible paths. Thus, we write dQ_{rev} as a total differential. The reversible work dW_{rev} between two neighboring states is also a total differential since the reversible work is the same for all reversible paths.

For a reversible process, the first law can be written as

$$dU = dQ_{rev} - dW_{rev} = T\, dS - p\, dV \qquad (7.3)$$

Here, we have written $dW_{rev} = p\, dV$ and $dQ_{rev} = T\, dS$. Comparing Eqs. 7.1 and 7.3, we note that for an irreversible process $\delta Q \neq T\, dS$ and $\delta W \neq pdV$. Since Eq. 7.3 is a relationship expressing the change of a state function U, it is valid for all processes.

If we were to define specific entropy as entropy per unit mass $\left(s = \dfrac{S}{m} \right)$ in a manner similar to specific internal energy u, Eq. 7.3 reduces to

$$du = T\, ds - pdv \qquad (7.4)$$

where v is the specific volume, that is, volume per unit mass. In the above, $u(s, v)$, that is, specific internal energy u is a function of specific entropy (s) and specific volume (v).

Although the relationship 7.4 is derived based on reversible process, that is, $dQ_{rev} = T\, dS$ and $dW_{rev} = p\, dV$, it is a relation between state properties for two neighboring states and is not process-specific. So if we have an irreversible process between two states, the relation

$$dU = \delta Q - \delta W$$

applies and so does

$$du = T\, ds - pdv$$

though $T\, ds$ is no longer δQ and pdv is not δW.

7.2.3 ENTHALPY

In a number of processes involving open systems, the state function enthalpy $H = U + pV$ is more convenient to use. The term "$p\,V$" can be thought of as the displacement work done on a system for it to occupy a volume "V" in an environment at a pressure "p". Thus, the enthalpy is a more appropriate state function in open system analysis involving mass transfer.

From definition of enthalpy, we can write Eq. 7.3 as

$$dH = TdS + Vdp \qquad (7.5)$$

Writing specific enthalpy (enthalpy per unit mass) as $h = \dfrac{H}{m}$, we can write the above equation as

$$dh = Tds + vdp \qquad (7.6)$$

which is the relationship between state properties.

7.2.4 HELMHOLTZ- AND GIBBS-FREE ENERGIES

Two other important state functions are the Helmholtz function "A" and the Gibbs-free energy "G". They are defined as follows:

$$A = U - TS \qquad (7.7)$$

and

$$G = H - TS \qquad (7.8)$$

respectively. The specific Helmholtz function $\left(a = \dfrac{A}{m} \right)$ and specific Gibbs-free energy $\left(g = \dfrac{G}{m} \right)$ from Eqs. 7.7 and 7.8 become

$$a = u - Ts \qquad (7.9)$$

and

$$g = h - Ts \qquad (7.10)$$

From the first law, we write for any system

$$\delta W = \delta Q - dU$$

and the second law gives $\delta Q \leq T\,dS$. Therefore, the work done is an isothermal process in which the system interacts with a single reservoir at the same temperature T

$$\delta W \leq -d(U - TS)$$

or

$$\delta W \leq -dA \qquad (7.11)$$

Thus, the work done is less than the decrease in the Helmholtz function for any isothermal process. The equality sign refers to a reversible process in which case, the work is maximum and equal to the decrease of the Helmholtz function.

For an isobaric and isothermal process, we can write

$$\delta W = \delta W' + pdV \leq -dA$$

where $\delta W'$ is the non-pdV work. The non-pdV work can be written as

$$\delta W' \leq -dA - pdV = -d(A + pV) \tag{7.12}$$

Since

$$A + pV = U - TS + pV = H - TS = G \tag{7.13}$$

we obtain the result

$$\delta W' \leq -dG \tag{7.14}$$

that is, the non-pdV work done by a system for an isobaric and isothermal process corresponds to the decrease in the Gibbs-free energy. The term free energy is therefore used for the Gibbs function to avoid any confusion. If there is no non-pdV work, that is, $\delta W' = 0$, then

$$0 \leq -dG \tag{7.15}$$

In other words, the change in Gibbs-free energy is negative or the system evolves in the direction of decreasing Gibbs-free energy. At equilibrium for which no work is possible $dG = 0$ or G is a minimum.

7.2.5 SUMMARY OF RELATIONSHIPS BETWEEN STATE PROPERTIES

Summarizing, the state functions are U, H, A and G and their change between two neighboring states are:

$$dU(S,V) = TdS - pdV \tag{7.16}$$

$$dH(S,p) = TdS + Vdp \tag{7.17}$$

$$dA(T,V) = -SdT - pdV \tag{7.18}$$

$$dG(T,p) = -SdT + Vdp \tag{7.19}$$

The above four state functions, U, H, A and G, can be written per unit mass, namely, u, h, a and g as

$$du(s,v) = Tds - pdv \tag{7.20}$$

$$dh(s,p) = Tds + vdp \tag{7.21}$$

$$da(T,v) = -sdT - pdv \tag{7.22}$$

$$dg(T,p) = -sdT + vdp \tag{7.23}$$

The state functions enthalpy, Helmholtz function and Gibbs-free energy could also be mathematically derived using the Legendre transformation as given in the next section.

7.3 DERIVATION OF STATE FUNCTIONS USING THE LEGENDRE TRANSFORM

Consider a function $z(x, y)$. We write

$$dz(x,y) = \left(\frac{\partial z}{\partial x}\right)_y dx + \left(\frac{\partial z}{\partial y}\right)_x dy = a_1 dx + a_2 dy$$

where $a_1 = \left(\frac{\partial z}{\partial x}\right)_y$ and $a_2 = \left(\frac{\partial z}{\partial y}\right)_x$

If we write

$$a_1 dx = d(a_1, x) - x da_1$$

then the above equation becomes

$$dz(x,y) = d(a_1 x) - x da_1 + a_2 dy$$

and rearranging the above expression, we get

$$d(z - a_1 x) = -x da_1 + a_2 dy$$

We can define a new function $\alpha_1 = z - a_1 x$ and obtain

$$d\alpha_1(a_1, y) = -x da_1 + a_2 dy \qquad (7.24)$$

Writing $d(a_2, y) = a_2 dy + y da_2$, Eq. 7.24 can be rearranged to read

$$d(\alpha_1 - a_2 y) = d\alpha_2(a_1, a_2) = -x da_1 - y da_2 \qquad (7.25)$$

$\alpha_2(a_1, a_2)$ is now a function of the new variables a_1 and a_2

The above manipulation is known as Legendre's transformation and permits definition of new state functions.

Applying Legendre's transformation to the first law, that is, Eq. 7.4

$$du = Tds - pdv = du(s,v)$$

we obtain

$$dh(s,p) = d(u + pv) = T\,ds + v\,dp$$

where the new function

$$h = u + pv$$

is the specific enthalpy.
Writing Eq. 7.4 as

$$du = T\,ds - p\,dv = d(T\,s) - s\,dT - pdv$$

we obtain

$$d(u - T\,s) = -s\,dT - p\,dv = da(T,v)$$

where

$$a(T,v) = u - T\,s$$

is the Helmholtz function.
Similarly, we write for enthalpy

$$dh = d(T\,s) - s\,dT + v\,dp$$

$$\text{or } d(h - T\,s) = -sdT + v\,dp = dg(T,p)$$

where

$$g(T,p) = h - T\,s$$

is the Gibbs-free energy.

7.4   MAXWELL'S RELATIONSHIPS FOR STATE VARIABLES

The derivatives of the state functions u, h, a and g are total differentials. Using the condition for exact differentials, a number of important relationships can be obtained. From Eq. 7.4 for specific internal energy,

$$du = T\,ds - p\,dv$$

we obtain

$$\left(\frac{\partial T}{\partial v}\right)_s = -\left(\frac{\partial p}{\partial s}\right)_T \tag{7.26}$$

Similarly from Eqs. 7.6, 7.22 and 7.23, for specific enthalpy, Helmholtz and Gibbs-free energy, we get

$$\left(\frac{\partial T}{\partial p}\right)_s = \left(\frac{\partial v}{\partial s}\right)_p \tag{7.27}$$

$$\left(\frac{\partial s}{\partial v}\right)_T = \left(\frac{\partial p}{\partial T}\right)_v \tag{7.28}$$

$$\left(\frac{\partial s}{\partial p}\right)_T = -\left(\frac{\partial v}{\partial T}\right)_p \tag{7.29}$$

The above expressions (Eqs. 7.26–7.29) are referred to as Maxwell's equations. They relate to the state variables p, v, T and s in terms of partial derivatives. They do not refer to a specific process, but relate to the various variables for a given equilibrium state. It may be noted that though entropy is a state function, it can also be used as a variable in the same way as internal energy is used as a state variable.

7.5 THERMODYNAMIC POTENTIALS AND FORCES

In Eq. 7.4, the internal energy is expressed as a function of entropy and volume, that is, $u(s,v)$. We can therefore write

$$du(s,v) = \left(\frac{\partial u}{\partial s}\right)_v ds + \left(\frac{\partial u}{\partial v}\right)_s dv \tag{7.30}$$

and equating the coefficients of ds and dv in the above with Eq. 7.4, we obtain

$$T = \left(\frac{\partial u}{\partial s}\right)_v \quad p = -\left(\frac{\partial u}{\partial v}\right)_s \tag{7.31}$$

Similarly from Eq. 7.6 for $h(s,p)$

$$dh(s,p) = \left(\frac{\partial h}{\partial s}\right)_p ds + \left(\frac{\partial h}{\partial p}\right)_s dp \tag{7.32}$$

and equating coefficients of ds and dp from Eq. 7.6

$$T = \left(\frac{\partial h}{\partial s}\right)_p \quad v = -\left(\frac{\partial h}{\partial p}\right)_s \tag{7.33}$$

In a similar way, we write for $da(T,v)$ and $dg(T,p)$

$$da(T,v) = \left(\frac{\partial a}{\partial T}\right)_v dT + \left(\frac{\partial a}{\partial v}\right)_T dv \tag{7.34}$$

$$dg(T,p) = \left(\frac{\partial g}{\partial T}\right)_p dT + \left(\frac{\partial g}{\partial p}\right)_T dp \qquad (7.35)$$

Equating coefficients of dT, dv and dT and dp, respectively, in Eqs. 7.34 and 7.35 and comparing with Eqs. 7.22 and 7.23 for da and dg, respectively, we obtain

$$s = -\left(\frac{\partial a}{\partial T}\right)_v \qquad p = -\left(\frac{\partial a}{\partial v}\right)_T \qquad (7.36)$$

and

$$s = -\left(\frac{\partial g}{\partial T}\right)_p \qquad v = \left(\frac{\partial g}{\partial p}\right)_T \qquad (7.37)$$

Equations 7.31, 7.33, 7.36 and 7.37 give the thermodynamic variables T, p, v and s in terms of gradients of state functions u, h, a and g. In analogy to mechanics, we can interpret the state functions u, h, a and g as thermodynamic potential functions and the state variables s, T, p and v as thermodynamic forces or properties.

7.6 DETERMINATION OF STATE FUNCTIONS

7.6.1 INTERNAL ENERGY

Equation 7.4 gives the internal energy $u(s,v)$ as functions of entropy s and v. However, entropy is not a directly measurable variable and it is more convenient to use the measurable variables p, v and T as independent variables. Choosing T and v as independent variables, that is, $u(T,v)$ we write

$$du(T,v) = \left(\frac{\partial u}{\partial T}\right)_v dT + \left(\frac{\partial u}{\partial v}\right)_T dv \qquad (7.38)$$

The specific heats are the heat capacities on a per unit mass (or mole) basis defined as specific heat at constant volume $c_v = \left(\frac{\partial u}{\partial T}\right)_v$ and specific heat at constant pressure $c_p = \left(\frac{\partial h}{\partial T}\right)_p$. Equation 7.38 therefore becomes

$$du(T,v) = c_v dT + \left(\frac{\partial u}{\partial v}\right)_T dv$$

Since $a = u - Ts$

$$\left(\frac{\partial u}{\partial v}\right)_T = \left(\frac{\partial a}{\partial v}\right)_T + T\left(\frac{\partial s}{\partial v}\right)_T$$

and from Eq. 7.36 and the Maxwell equation (7.28), we obtain

$$\left(\frac{\partial u}{\partial v}\right)_T = -p + T\left(\frac{\partial p}{\partial T}\right)_v \tag{7.39}$$

Thus,

$$du(T,v) = c_v dT + \left[T\left(\frac{\partial p}{\partial T}\right)_v - p\right]dv \tag{7.40}$$

$\left(\frac{\partial p}{\partial T}\right)_v$ in the above equation can be evaluated from the equation of state $f(p,v,T) = 0$. Hence the change in internal energy can be expressed in terms of measurable quantities, that is, specific heat, thermodynamic properties and equation of state. As an example, for an ideal gas where $pv = RT$, Eq. 7.39 indicates that

$$\left(\frac{\partial u}{\partial v}\right)_T = -p + T\left(\frac{R}{v}\right) = 0$$

Hence, u is a function of T alone for an ideal gas, that is,

$$du = c_v dT \tag{7.41}$$

If p,T are chosen as independent variables, then

$$du(p,T) = \left(\frac{\partial u}{\partial T}\right)_p dT + \left(\frac{\partial u}{\partial p}\right)_T dp \tag{7.42}$$

From definition of specific enthalpy

$$h = u + pv$$

we write

$$\left(\frac{\partial u}{\partial T}\right)_p = \left(\frac{\partial h}{\partial T}\right)_p - p\left(\frac{\partial v}{\partial T}\right)_p = c_p - p\left(\frac{\partial v}{\partial T}\right)_p \tag{7.43}$$

With $a = u - Ts$, we get

$$\left(\frac{\partial u}{\partial p}\right)_T = \left(\frac{\partial a}{\partial p}\right)_T + T\left(\frac{\partial s}{\partial p}\right)_T \tag{7.44}$$

and using Eq. 7.36, we get

$$\left(\frac{\partial a}{\partial p}\right)_T = \left(\frac{\partial a}{\partial v}\right)_T \left(\frac{\partial v}{\partial p}\right)_T = -p\left(\frac{\partial v}{\partial p}\right)_T \qquad (7.45)$$

Substituting the values of the partial derivatives from Eqs. 7.43 and 7.44 in Eq. 7.42 with Maxwell relation (7.29), we write

$$du(p,\mathrm{T}) = \left[c_p - p\left(\frac{\partial v}{\partial T}\right)_p\right]dT + \left[-p\left(\frac{\partial v}{\partial p}\right)_T - T\left(\frac{\partial v}{\partial T}\right)_p\right]dp \qquad (7.46)$$

Thus, the change in $u(p,T)$ for the corresponding change in temperature and pressure is expressed in terms of the specific heat c_p, the isothermal compressibility $\left(\frac{\partial v}{\partial p}\right)_T$ and the volume expansion term $\left(\frac{\partial v}{\partial T}\right)_p$.

In terms of p,v as independent variables, we write

$$du(p,v) = \left(\frac{\partial u}{\partial p}\right)_v dp + \left(\frac{\partial u}{\partial v}\right)_p dv$$

Since

$$\left(\frac{\partial u}{\partial p}\right)_v = \left(\frac{\partial u}{\partial T}\right)_v \left(\frac{\partial T}{\partial p}\right)_v = c_v\left(\frac{\partial T}{\partial p}\right)_v$$

Also with $h = u + pv$

$$\left(\frac{\partial u}{\partial p}\right)_p = \left(\frac{\partial h}{\partial v}\right)_p - p = \left(\frac{\partial h}{\partial T}\right)_p\left(\frac{\partial T}{\partial v}\right) - p = c_p\left(\frac{\partial T}{\partial v}\right)_p - p$$

Hence,

$$du(p,v) = \left[c_v\left(\frac{\partial T}{\partial p}\right)_v\right]dp + \left[c_p\left(\frac{\partial T}{\partial v}\right)_p - p\right]dv \qquad (7.47)$$

For an ideal gas where, $pv = RT$, $\left(\frac{\partial T}{\partial p}\right)_v = \frac{v}{R}$ and $\left(\frac{\partial T}{\partial v}\right)_p = \frac{p}{R}$ giving

$$du(p,v) = \frac{c_v}{R}vdp + p\left(\frac{c_p}{R} - 1\right)dv$$

With $c_p - c_v = R$ for an ideal gas, the above becomes

$$du(p,v) = \frac{c_v}{R}(vdp + pdv) = \frac{c_v}{R}d(pv) = \frac{c_v}{R}(RdT) = c_v dT$$

We therefore obtain the same result $du = c_v dT$ for an ideal gas.

Internal energy changes can thus be obtained from Eqs. 7.40, 7.46 and 7.47 as functions of the changes in (v, T), (p, T) and (p, v) respectively. They all reduce to $du = c_v dT$ for an ideal gas.

We can also show that the internal energy for an ideal gas is a function of temperature T and not v through the following alternate procedure.

We have for internal energy:

$$du(T,v) = \left(\frac{\partial u}{\partial T}\right)_v dT + \left(\frac{\partial u}{\partial v}\right)_T dv$$

and from the first law

$$\delta Q_{rev} = du + \delta W_{rev} = du + pdv$$

or

$$Tds = du + pdv \text{ giving}$$

$$ds = \frac{1}{T} du + \frac{p}{T} dv$$

The above can further be written as

$$ds = \frac{1}{T}\left[\left(\frac{\partial u}{\partial v}\right)_T dv + \left(\frac{\partial u}{\partial T}\right)_v dT\right] + \frac{p}{T} dv$$

Simplifying we get

$$ds = \left[\frac{1}{T}\left(\frac{\partial u}{\partial v}\right)_T + \frac{p}{T}\right] dv + \frac{1}{T}\left(\frac{\partial u}{\partial T}\right)_v dT$$

But $ds(v,T) = \left(\frac{\partial s}{\partial v}\right)_T dv + \left(\frac{\partial s}{\partial T}\right)_v dT$

and $\frac{\partial}{\partial v}\bigg)_T \left(\frac{\partial s}{\partial T}\right)_v = \frac{\partial}{\partial T}\bigg)_v \left(\frac{\partial s}{\partial v}\right)_T$

For an ideal gas therefore

$$\frac{\partial}{\partial v}\bigg)_T \left[\frac{1}{T}\left(\frac{\partial u}{\partial T}\right)_v\right] = \frac{\partial}{\partial T}\bigg)_v \left[\frac{1}{T}\left(\frac{\partial u}{\partial v}\right)_T + \frac{R}{v}\right]$$

Hence, $\dfrac{1}{T}\dfrac{\partial^2 u}{\partial v \partial T} = \dfrac{1}{T}\dfrac{\partial^2 u}{\partial T \partial v} - \dfrac{1}{T^2}\left(\dfrac{\partial u}{\partial v}\right)_T$

This gives

$$\left(\frac{\partial u}{\partial v}\right)_T = 0 \tag{7.48}$$

so that $u \neq u(v)$ but $u = u(T)$ alone for an ideal gas.

7.6.2 Enthalpy

If it is desired to express the enthalpy as a function of T, p, that is, $h(T, p)$ we write

$$dh(T, p) = \left(\frac{\partial h}{\partial T}\right)_p dT + \left(\frac{\partial h}{\partial p}\right)_T dp$$

But

$$\left(\frac{\partial h}{\partial T}\right)_p = c_p$$

and using the Gibbs function

$$g = h - Ts$$

we write

$$\left(\frac{\partial h}{\partial p}\right)_T = \left(\frac{\partial g}{\partial p}\right)_T + T\left(\frac{\partial s}{\partial p}\right)_T$$

From Eq. 7.37 and Maxwell's equation (7.29), we have $\left(\frac{\partial g}{\partial p}\right)_T = v; \quad \left(\frac{\partial s}{\partial p}\right)_T = -\left(\frac{\partial v}{\partial T}\right)_p$

and therefore

$$\left(\frac{\partial h}{\partial p}\right)_T = v - T\left(\frac{\partial v}{\partial T}\right)_p$$

Hence,

$$dh(p, T) = c_p dT + \left(v - T\left(\frac{\partial v}{\partial T}\right)_p\right) dp \tag{7.49}$$

For an ideal gas $\left(\frac{\partial v}{\partial T}\right)_p = \frac{R}{p}$ and therefore

$$dh(p, T) = dh(T) = c_p dT$$

that is, enthalpy of an ideal gas is a function of temperature only.

For specific enthalpy expressed as a function of temperature and specific volume $h(T,v)$, we first write

$$dh(T,v) = \left(\frac{\partial h}{\partial T}\right)_v dT + \left(\frac{\partial h}{\partial v}\right)_T dv$$

Since $h = u + pv$

$$\left(\frac{\partial h}{\partial T}\right)_v = \left(\frac{\partial u}{\partial T}\right)_v + v\left(\frac{\partial p}{\partial T}\right)_V = c_v + v\left(\frac{\partial p}{\partial T}\right)_v$$

Using the Gibbs-free energy, $g = h - Ts$

we can express $\left(\frac{\partial h}{\partial v}\right)_T = \left(\frac{\partial g}{\partial v}\right)_T + T\left(\frac{\partial s}{\partial v}\right)_T = \left(\frac{\partial g}{\partial p}\right)_T\left(\frac{\partial p}{\partial v}\right)_T + T\left(\frac{\partial p}{\partial T}\right)_v$

Here, we have used Maxwell equation (7.28), viz. $\left(\frac{\partial s}{\partial v}\right)_T = \left(\frac{\partial p}{\partial T}\right)_v$

Further using Eq. 7.37, the above becomes

$$\left(\frac{\partial h}{\partial v}\right)_T = v\left(\frac{\partial p}{\partial v}\right)_T + T\left(\frac{\partial p}{\partial T}\right)_v$$

and we obtain

$$dh(T,v) = \left(c_v + v\left(\frac{\partial p}{\partial T}\right)_v\right)dT + \left(v\left(\frac{\partial p}{\partial v}\right)_T + T\left(\frac{\partial p}{\partial T}\right)_v\right)dv \qquad (7.50)$$

For an ideal gas where $pv = RT$

$\left(\frac{\partial p}{\partial T}\right)_v = \frac{R}{v}$, $\left(\frac{\partial p}{\partial v}\right)_T = -\frac{RT}{v^2}$, and therefore the above equation becomes

$$dh(T,v) = (c_v + R)dT + \left(-p + \frac{RT}{v}\right)dv = (c_v + R)dT = c_p dT$$

the same result as obtained previously indicating that $h(T)$ only.

Similarly for, $h(p,v)$ we write

$$dh(p,v) = \left(\frac{\partial h}{\partial p}\right)_v dp + \left(\frac{\partial h}{\partial v}\right)_p dv$$

where

$$\left(\frac{\partial h}{\partial v}\right)_p = \left(\frac{\partial h}{\partial T}\right)_p\left(\frac{\partial T}{\partial v}\right)_p = c_p\left(\frac{\partial T}{\partial v}\right)_p$$

and for $\left(\dfrac{\partial h}{\partial p}\right)_v$ we obtain using $h = u + pv$

$$\left(\frac{\partial h}{\partial p}\right)_v = \left(\frac{\partial u}{\partial p}\right)_v + v = \left(\frac{\partial u}{\partial T}\right)_v\left(\frac{\partial T}{\partial p}\right)_v + v$$

Hence,

$$dh(p,v) = c_p\left(\frac{\partial T}{\partial v}\right)_p dv + \left(c_v\left(\frac{\partial T}{\partial p}\right)_v + v\right)dp \qquad (7.51)$$

For an ideal gas where $pv = RT$ and $\left(\dfrac{\partial T}{\partial v}\right)_p = \dfrac{p}{R}$, $\cdot\left(\dfrac{\partial T}{\partial p}\right)_v = \dfrac{v}{R}$, we obtain

$$dh(p,v) = \left(c_p\frac{p}{R}\right)dv + \left(c_v\frac{v}{R} + v\right)dp$$

$$= \frac{c_p}{R}(pdv + vdp) = \frac{c_p}{R}d(pv) = c_p dT$$

Enthalpy changes can thus be obtained from Eqs. 7.49, 7.50 and 7.51 as functions of changes in (T, v), (p, T) and (p, v) respectively. They all reduce to $dh = c_p dT$ for an ideal gas.

7.6.3 ENTROPY

Entropy is defined as

$$ds = \frac{dQ_{\text{rev}}}{T}$$

Thus to determine the specific entropy change between two states, one chooses a reversible path to connect two states and compute the reversible heat transfer per unit mass of the system. However, analytical expressions can also be obtained to determine the entropy change. If T and v are the desired independent variables, we first write

$$ds(T,v) = \left(\frac{\partial s}{\partial T}\right)_v dT + \left(\frac{\partial s}{\partial v}\right)_T dv$$

We also note

$$\left(\frac{\partial s}{\partial T}\right)_v = \left(\frac{\partial s}{\partial u}\right)_v\left(\frac{\partial u}{\partial T}\right)_v = \frac{1}{T}c_v$$

Maxwell's equation (7.28) gives $\left(\dfrac{\partial s}{\partial v}\right)_T = \left(\dfrac{\partial p}{\partial T}\right)_v$.

Hence,

$$ds(T,v) = c_v \frac{dT}{T} + \left(\frac{\partial p}{\partial T}\right)_v dv \tag{7.52}$$

For $s(p,T)$, we write

$$ds(p,T) = \left(\frac{\partial s}{\partial p}\right)_T dp + \left(\frac{\partial s}{\partial T}\right)_p dT$$

and using the relation

$$\left(\frac{\partial s}{\partial T}\right)_p = \left(\frac{\partial s}{\partial h}\right)_p \left(\frac{\partial h}{\partial T}\right)_p = \frac{1}{T} c_p$$

and also the Maxwell equation (7.29) which gives $\left(\frac{\partial s}{\partial p}\right)_T = -\left(\frac{\partial v}{\partial T}\right)_p$, we get

$$ds(T,p) = c_p \frac{dT}{T} - \left(\frac{\partial v}{\partial T}\right)_p dp \tag{7.53}$$

For $s(p,v)$, we first write

$$ds(p,v) = \left(\frac{\partial s}{\partial p}\right)_v dp + \left(\frac{\partial s}{\partial v}\right)_p dv$$

The differential $\left(\frac{\partial s}{\partial p}\right)_v$ can be written as

$$\left(\frac{\partial s}{\partial p}\right)_v = \left(\frac{\partial s}{\partial u}\right)_v \left(\frac{\partial u}{\partial T}\right)_v \left(\frac{\partial T}{\partial p}\right)_v = \frac{1}{T} c_v \left(\frac{\partial T}{\partial p}\right)_v$$

Similarly for $\left(\frac{\partial s}{\partial v}\right)_p$, we have

$$\left(\frac{\partial s}{\partial v}\right)_p = \left(\frac{\partial s}{\partial h}\right)_p \left(\frac{\partial h}{\partial T}\right)_p \left(\frac{\partial T}{\partial v}\right)_p = \frac{1}{T} c_p \left(\frac{\partial T}{\partial v}\right)_p$$

Thus, we obtain

$$ds(p,v) = \frac{c_v}{T} \left(\frac{\partial T}{\partial p}\right)_v dp + \frac{c_p}{T} \left(\frac{\partial T}{\partial v}\right)_p dv \tag{7.54}$$

Equations 7.52–7.54 give entropy changes as function of changes in (T, v), (T, p) and (p, v) respectively

For an ideal gas where $pv = RT$, $\left(\dfrac{\partial p}{\partial T}\right) = \dfrac{R}{v}$, $\left(\dfrac{\partial v}{\partial T}\right) = \dfrac{R}{p}$, Eqs. 7.52–7.54 give

$$ds(T,v) = c_v \frac{dT}{T} + \frac{R}{v} dv \tag{7.55}$$

$$ds(T,p) = c_p \frac{dT}{T} - \frac{R}{p} dp \tag{7.56}$$

$$ds(p,v) = c_v \frac{dp}{p} + c_p \frac{dv}{v} \tag{7.57}$$

If specific heats $c_p(T)$ and $c_v(T)$ are given, we can readily integrate the above equations for an ideal gas to determine the change of entropy. For constant values of c_p and c_v, the above equations integrate to yield

$$\frac{\Delta s(T,v)}{c_v} = \ln\left(\frac{T_2}{T_1}\right)\left(\frac{v_2}{v_1}\right)^{\gamma-1} \tag{7.58}$$

$$\frac{\Delta s(T,p)}{c_p} = \ln\left(\frac{T_2}{T_1}\right)\left(\frac{p_2}{p_1}\right)^{\frac{\gamma-1}{\gamma}} \tag{7.59}$$

$$\frac{\Delta s(p,v)}{c_v} = \ln\left(\frac{p_2}{p_1}\right)\left(\frac{v_2}{v_1}\right)^{\gamma} \tag{7.60}$$

7.7 THERMODYNAMIC FUNCTIONS FOR DENSE GASES

The equation of state for dense gases is more complex. The p, v and T data may be obtained from empirical equations of state or approximately from the compressibility chart. As an example, consider the van der Waal's equation of state given below:

$$\left(p + \frac{a}{\tilde{v}^2}\right)(\tilde{v} - b) = R_0 T \tag{7.61}$$

In the above expression, "a" and "b" are constants fitted for the experimental p, v and T data in the region of interest. The specific volume is expressed on molar basis, that is,

$$\tilde{v} = \frac{V}{n}$$

and R_0 is the universal gas constant.

From Eq. 7.61, we write

$$p = \frac{R_0 T}{\tilde{v} - b} - \frac{a}{\tilde{v}^2} \tag{7.62}$$

and thus

$$\left(\frac{\partial p}{\partial T} \right)_v = \frac{R_0}{\tilde{v} - b} \tag{7.63}$$

Writing Eq. 7.40 for $du(T, v)$ on a molar basis, we have

$$d\tilde{u}(T, \tilde{v}) = \tilde{c}_v dT + \left(T \left(\frac{\partial p}{\partial T} \right)_{\tilde{v}} - p \right) d\tilde{v}$$

and using Eqs. 7.63 and 7.62, we get

$$d\tilde{u}(T, \tilde{v}) = \tilde{c}_v dT + \frac{a}{\tilde{v}^2} d\tilde{v} \tag{7.64}$$

which integrates to yield

$$\Delta \tilde{u}(T, \tilde{v}) = \int_{T_1}^{T_2} \tilde{c}_v dT + \int_{\tilde{v}_1}^{\tilde{v}_2} \frac{a}{\tilde{v}^2} d\tilde{v}$$

$$= \int_{T_1}^{T_2} \tilde{c}_v dT - a \left(\frac{1}{\tilde{v}_2} - \frac{1}{\tilde{v}_1} \right) \tag{7.65}$$

Similarly from Eq. 7.49, we express $dh(p, T)$ on a molar basis as

$$d\tilde{h}(T, p) = \tilde{c}_p dT + \left(\tilde{v} - T \left(\frac{\partial \tilde{v}}{\partial T} \right)_p \right) dp \tag{7.66}$$

We use the cyclic rule for variables $\tilde{v}, T, p$ to obtain $\left(\frac{\partial \tilde{v}}{\partial T} \right)_p$, that is,

$$\left(\frac{\partial \tilde{v}}{\partial T} \right)_p \left(\frac{\partial T}{\partial p} \right)_{\tilde{v}} \left(\frac{\partial p}{\partial \tilde{v}} \right)_T = -1$$

$$\left(\frac{\partial \tilde{v}}{\partial T} \right)_p = \frac{-1}{\left(\frac{\partial T}{\partial p} \right)_{\tilde{v}} \left(\frac{\partial p}{\partial \tilde{v}} \right)_T} = -\frac{\left(\frac{\partial p}{\partial T} \right)_{\tilde{v}}}{\left(\frac{\partial p}{\partial \tilde{v}} \right)_T} \tag{7.67}$$

From the equation $p = \dfrac{R_0 T}{\tilde{v} - b} - \dfrac{a}{\tilde{v}^2}$, we obtain

$$\left(\frac{\partial p}{\partial \tilde{v}}\right)_T = -\frac{R_0 T}{(\tilde{v} - b)^2} + \frac{2a}{\tilde{v}^3} = \frac{2a(\tilde{v} - b)^2 - R_0 T \tilde{v}^3}{\tilde{v}^3 (\tilde{v} - b)^2} \qquad (7.68)$$

Using the above and $\left(\dfrac{\partial p}{\partial T}\right)_{\tilde{v}} = \dfrac{R_0}{\tilde{v} - b}$, we then write

$$\left(\frac{\partial \tilde{v}}{\partial T}\right)_p = \frac{\left(\dfrac{\partial p}{\partial T}\right)_{\tilde{v}}}{\left(\dfrac{\partial p}{\partial \tilde{v}}\right)_T} = \frac{\dfrac{-R_0}{(\tilde{v} - b)}}{\dfrac{2a(\tilde{v} - b)^2 - R_0 T \tilde{v}^3}{\tilde{v}^3 (\tilde{v} - b)^2}} = \frac{R_0 \tilde{v}^3 (\tilde{v} - b)}{R_0 T \tilde{v}^3 - 2a(\tilde{v} - b)^2} \qquad (7.69)$$

and using Eq. 7.69 in Eq. 7.66, we get

$$d\tilde{h}(T, p) = \tilde{c}_p dT + \left(\tilde{v} - \frac{R_0 T \tilde{v}^3 (\tilde{v} - b)}{R_0 T \tilde{v}^3 - 2a(\tilde{v} - b)^2} \right) dp \qquad (7.70)$$

Alternatively, a simpler expression can be obtained using the definition of enthalpy, that is,

$$\tilde{h} = \tilde{u} + p\tilde{v}$$

to give

$$d\tilde{h} = d\tilde{u} + d(p\tilde{v})$$

Substituting the value of change of the specific molar internal energy from Eq. 7.64, that is,

$$d\tilde{u}(T, \tilde{v}) = \tilde{c}_v dT + \frac{a}{\tilde{v}^2 d\tilde{v}}$$

we obtain

$$d\tilde{h} = \tilde{c}_v dT + \frac{a}{\tilde{v}^2} d\tilde{v} + d(p\tilde{v})$$

which integrates to yield

$$\Delta \tilde{h} = \int_{T_1}^{T_2} \tilde{c}_v \, dT - a\left(\frac{1}{\tilde{v}_2} - \frac{1}{\tilde{v}_1} \right) + (p_2 \tilde{v}_2 - p_1 \tilde{v}_1) \qquad (7.71)$$

The van der Waal's equation (7.61) can be used for determining the product $p\tilde{v}$.
To determine the entropy $s(T, v)$, we use Eq. 7.52, which gives

$$ds(T,v) = c_v \frac{dT}{T} + \left(\frac{\partial p}{\partial T}\right)_v dv$$

and using the value of

$$\left(\frac{\partial p}{\partial T}\right)_{\tilde{v}} = \frac{R_0}{\tilde{v} - b}$$

from the van der Waals equation, we get the entropy change per mole as

$$d\tilde{s} = \tilde{c}_{\tilde{v}} \frac{dT}{T} + \left(\frac{R_0}{\tilde{v} - b}\right) d\tilde{v}$$

which integrates to yield

$$\Delta\tilde{s} = \int_{T_1}^{T_2} \tilde{c}_{\tilde{v}} \frac{dT}{T} + R_0 \ln \frac{(\tilde{v}_2 - b)}{(\tilde{v}_1 - b)} \qquad (7.72)$$

For low pressures where the molar volume tends to infinity, the above reduces to the entropy of an ideal gas. Similarly, it may be noted that the changes in internal energy

$$\Delta\tilde{u}(T, \tilde{v})$$

and enthalpy

$$\Delta\tilde{h}(T, \tilde{v})$$

given by Eqs. 7.65 and 7.71 reduce for an ideal gas to

$$\int_{T_1}^{T_2} \tilde{C}_v \, dT$$

and

$$\int_{T_1}^{T_2} \tilde{c}_v \, dT + (p_2\tilde{v}_2 - p_1\tilde{v}_1) = \int_{T_1}^{T_2} \tilde{c}_p \, dT$$

respectively.

7.8   GENERALIZED ENTHALPY AND ENTROPY CHARTS

For dense gases, the compressibility chart provides an approximate estimate of the p, v and T data from the knowledge of the critical pressure and temperature of the substance. Enthalpy and entropy functions for dense gases can be estimated from the so-called generalized enthalpy and entropy charts. Enthalpy as a function of T, p is given by Eq. 7.49 as

$$d\tilde{h}(T,p) = \tilde{c}_p dT + \left(\tilde{v} - T\left(\frac{\partial \tilde{v}}{\partial T} \right)_p \right) dp$$

which for an ideal gas where

$$p\tilde{v} = R_0 T$$

gives

$$d\tilde{h}(T,p) = \tilde{c}_p dT$$

and enthalpy is a function only of the temperature for an ideal gas. Departure from an ideal gas behavior is given by the second term on the right hand side of the expression for $d\tilde{h}(T,p)$, that is,

$$\left(\tilde{v} - T\left(\frac{\partial \tilde{v}}{\partial T} \right)_p \right) dp$$

Integrating $d\tilde{h}(T,p)$ along an isotherm, we write

$$\tilde{h}(T,p) - \tilde{h}(T,p_0) = \int_{p_0}^{p} \left[\tilde{v} - T\left(\frac{\partial \tilde{v}}{\partial T} \right)_p \right] dp \qquad (7.73)$$

If we let p_0 tend to zero, we have an ideal gas and

$$\tilde{h}(T,p_0) \rightarrow \tilde{h}^*(T)$$

where

$$\tilde{h}^*(T)$$

denotes the enthalpy of an ideal gas. Knowledge of an explicit equation of state permits the integral in Eq. 7.73 to be evaluated.

However, the compressibility chart can also be used more conveniently. Using the compressibility factor "Z", we write

$$\tilde{v} = \frac{ZR_0 T}{p}$$

and

$$\left(\frac{\partial \tilde{v}}{\partial T}\right)_p = \frac{ZR_0}{p} + \frac{R_0 T}{p}\left(\frac{\partial Z}{\partial T}\right)_p$$

Using the above, Eq. 7.73 becomes

$$\tilde{h}(T,p) - \tilde{h}^*(T) = \int\limits_{p_o \to 0}^{p} -\left[\frac{R_0 T^2}{p}\left(\frac{\partial Z}{\partial T}\right)_p\right]dp$$

where $h^*(T)$ denotes the enthalpy of an ideal gas where $p_o \to 0$. Normalizing the pressure p and temperature T with respect to the critical pressure p_c and temperature T_c to give relative pressure p_R and relative temperature T_R, the above equation can be rearranged to read

$$\frac{\tilde{h}^*(T) - \tilde{h}(T,p)}{R_0 T_c} = \int\limits_{p_0 \to 0}^{p_R} T_R^2 \left(\frac{\partial Z}{\partial T_R}\right)_{p_R} d\ln p_R \qquad (7.74)$$

The above is referred to as the enthalpy departure function on molar basis. The integral is carried out along an isotherm $T_R = $ constant from given (p, v, T) data. A plot of

$$\frac{\tilde{h}^*(T) - \tilde{h}(T,p_R)}{R_0 T_c}$$

against p_R for constant T_R is referred to as generalized enthalpy chart and is shown in Figure 7.1.

For enthalpy change between two states, we write

$$\int\limits_{\tilde{h}_1(T_1,p_1)}^{\tilde{h}_2(T_2,p_2)} d\tilde{h} = \tilde{h}_2(T_2,p_2) - \tilde{h}_1(T_1,p_1) = -\left[\tilde{h}_2^*(T_2) - \tilde{h}_2(T_2,p_2)\right]$$

$$+\left[\tilde{h}_1^*(T_1) - \tilde{h}_1(T_1,p_1)\right] + \left[\tilde{h}^*(T_2) - \tilde{h}^*(T_1)\right]$$

$$= -\left[\frac{\tilde{h}_2^*(T_2) - \tilde{h}_2(T_2,p_2)}{R_0 T_c}\right]R_0 T_c + \left[\tilde{h}_2^*(T_2) - \tilde{h}_1^*(T_1)\right]$$

$$+\left[\frac{\tilde{h}_1^*(T_1) - \tilde{h}_1(T,p_1)}{R_0 T_c}\right]R_0 T_c \qquad (7.75)$$

Given (T, p), the departure function can be obtained from the generalized enthalpy chart (Figure 7.1). The enthalpy change for an ideal gas

$$\tilde{h}_2^*(T_2) - \tilde{h}_1^*(T_1)$$

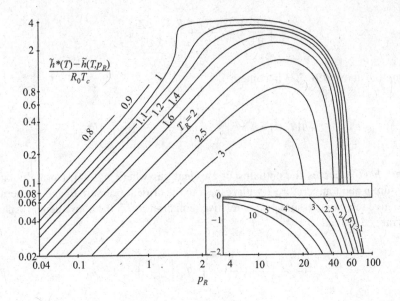

FIGURE 7.1   Generalized enthalpy chart.

can be determined using ideal gas tables or from integrating

$$\tilde{c}_p^{\,*}\,dT$$

if the specific heat $\tilde{c}_p^{\,*}(T)$ of ideal gas is known. Thus, Eq. 7.75 can be used to obtain

$$\tilde{h}_2(T_2, p_2) - \tilde{h}_1(T_1, p_1)$$

The specific molar entropy from Eq. 7.53 is given by

$$d\tilde{s}(T, p) = \tilde{c}_p \frac{dT}{T} - \left(\frac{\partial \tilde{v}}{\partial T}\right)_p dp \qquad (7.76)$$

For the entropy of an ideal gas for which

$$\left(\frac{\partial \tilde{v}}{\partial T}\right)_p = \frac{R_0}{p}$$

we have

$$d\tilde{s}^{*}(T, p) = \tilde{c}_p \frac{dT}{T} - \frac{R_0}{p} dp$$

Unlike the enthalpy, the entropy for an ideal gas depends on both p and T. Integrating the above along an isotherm between pressures p_0 and p, we have

$$\int\limits_{\tilde{s}^*(T,p_0)}^{\tilde{s}^*(T,p)} d\tilde{s}^* = \tilde{s}^*(T,p) - \tilde{s}^*(T,p_0) = -\int\limits_{p0}^{p} \frac{R_0}{p} dp$$

Subtracting the two expressions above for ideal and dense gases gives

$$\left[\tilde{s}^*(T,p) - \tilde{s}^*(T,p_0)\right] - \left[\tilde{s}(T,p) - \tilde{s}(T,p_0)\right] = \int\limits_{p0}^{p}\left[\left(\frac{\partial \tilde{v}}{\partial T}\right)_p - \frac{R_0}{p}\right]dp$$

where p_0 is sufficiently low for the ideal gas equation to be valid.
Using compressibility factor Z, the above equation can be written as

$$\left[\tilde{s}^*(T,p) - \tilde{s}(T,p)\right] - \left[\tilde{s}^*(T,p_0) - \tilde{s}(T,p_0)\right] = \int\limits_{p0}^{p}\left((z-1)\frac{R_0}{p} + \frac{R_0 T}{p}\left(\frac{\partial Z}{\partial T}\right)_p\right)dp$$

In the limit when

$$p_0 \to 0, \tilde{s}(T,p_0) \to \tilde{s}^*(T,p_0)$$

Normalizing with respect to critical pressure p_c and critical temperature T_c, the above expression becomes

$$\frac{\tilde{s}^*(T,p) - \tilde{s}(T,p)}{R_0} = \int\limits_{p_0 \to 0}^{P_R} \left((z-1)d \ln P_R\right) + \int\limits_{p_0 \to 0}^{P_R} \frac{T_R}{P_R}\left(\frac{\partial Z}{\partial T}\right)_{P_R} dP_R \qquad (7.77)$$

Equation 7.77 gives the departure of the entropy at (p, T) from the ideal gas value at the same (p, T). The term

$$\frac{\tilde{s}^*(T,p) - \tilde{s}(T,p)}{R_0}$$

is referred to as the entropy departure function and is plotted against the reduced pressure $p_R = p/p_c$ for various reduced temperature $T_R = T/T_c$. Figure 7.2 shows the entropy departure function.
From Eq. 7.74, where

$$\frac{\tilde{h}^*(T) - \tilde{h}(T,p)}{R_0 T_c} = \int\limits_{P_0 \to 0}^{P_R} T_R^2\left(\frac{\partial Z}{\partial T_R}\right)_{P_R} d \ln p_R$$

we can also write the entropy departure function using Eq. 7.77 as

$$\frac{\tilde{s}^*(T,p)-\tilde{s}(T,p)}{R_0} = \frac{\tilde{h}^*(T)-h(T,p)}{R_0 T_c T_R} + \int_{P_0 \to 0}^{P_R} (z-1)d\ln P_R \qquad (7.78)$$

For the entropy change between two states p_1, T_1 and p_2, T_2, we can write

$$\tilde{s}_2(p_2,T_2)-\tilde{s}_1(p_1,T_1) = -\left[\frac{\tilde{s}_2^*(T_2,p_2)-\tilde{s}_2(T_2,p_2)}{R_0}\right]R_0 + \left[\tilde{s}_2^*(T_2,p_2)-\tilde{s}_1^*(T_1,p_1)\right]$$

$$+\left[\frac{\tilde{s}_1^*(T_1,p_1)-\tilde{s}_1(T_1,p_1)}{R_0}\right]R_0$$

The entropy departure function in the above expression

$$\left[\frac{\tilde{s}^*(T,p)-\tilde{s}(T,p)}{R_0}\right]$$

can be determined from the generalized entropy chart (Figure 7.2).

The entropy change for an ideal gas

$$\left[\tilde{s}_2^*(T_2,p_2)-\tilde{s}_1^*(T_1,p_1)\right]$$

can be obtained from the ideal gas tables or from integration if the variation of the molar specific heat with temperature is given. Thus the entropy change between the two states (T_1, p_1) and (T_2, p_2) is determined.

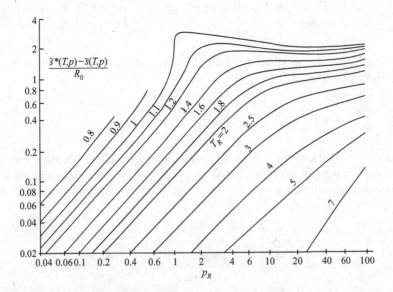

FIGURE 7.2   Entropy departure function.

8 Thermodynamic Coefficients and Specific Heats

8.1 THERMODYNAMIC COEFFICIENTS

The thermodynamic state variables such as pressure p, specific volume v and temperature T can be measured directly. Fitting the experimental p, v and T data gives the equation of state $f(p,v,T) = 0$ for the substance. The thermodynamic coefficients defined in the following are determined from the p, v and T data, while the specific heats are obtained from the dependence of the state functions internal energy and enthalpy on temperature, pressure and volume. In this chapter, we discuss the thermodynamic coefficients and specific heats and how to determine them.

8.1.1 COEFFICIENT OF VOLUME EXPANSION

The slope of a constant pressure curve at a point on the $v - T$ diagram gives $\left(\dfrac{\partial v}{\partial T}\right)_p$ and normalizing with v, we define the coefficient of volume expansion β as

$$\beta = \frac{1}{v}\left(\frac{\partial v}{\partial T}\right)_p \tag{8.1}$$

Values of β can be obtained and tabulated for a range of pressure and temperature.

8.1.2 ISOTHERMAL AND ISENTROPIC COMPRESSIBILITY

The slope of a constant temperature line on a $p - v$ plot gives isothermal compressibility K_T, that is,

$$K_T = -\frac{1}{v}\left(\frac{\partial v}{\partial p}\right)_T \tag{8.2}$$

The negative sign is used since for most substances the volume decreases with increase of pressure and therefore permits positive values for the isothermal compressibility K_T. Compressibility under adiabatic (or isentropic) conditions gives isentropic compressibility

DOI: 10.1201/9781003224044-8

$$K_s = -\frac{1}{v}\left(\frac{\partial v}{\partial P}\right)_s \qquad (8.3)$$

κ_s is essentially the slope of an isentrope on the $p-v$ diagram. The subscript "s" in Eq. 8.3 denotes volume change with pressure at constant entropy, that is, corresponding to a reversible adiabatic process.

8.1.3 PRESSURE COEFFICIENT

A pressure coefficient α is defined as change in pressure with temperature under constant volume, that is,

$$\alpha = \frac{1}{p}\left(\frac{\partial p}{\partial T}\right)_v \qquad (8.4)$$

This is the slope of a constant volume curve on a $p-T$ plot.

8.1.4 RELATIONSHIPS AMONG THE COEFFICIENTS

From calculus, a relationship between α, β and K_T can be obtained. Using the cyclic rule for the three variables p, v and T, we write

$$\left(\frac{\partial v}{\partial p}\right)_T \left(\frac{\partial p}{\partial T}\right)_v \left(\frac{\partial T}{\partial v}\right)_p = -1$$

Hence,

$$p\frac{k_T \alpha}{\beta} = 1 \qquad (8.5)$$

Equation 8.5 permits any of the three coefficients, α, β and K_T, to be determined if two of them are known.

The isentropic compressibility K_s, given by Eq. 8.3, is difficult to determine experimentally. However, K_s is related to the sound speed "a", which can be measured more easily. The sound speed is defined as

$$a^2 = \left(\frac{\partial p}{\partial \rho}\right)_s = -v^2\left(\frac{\partial p}{\partial v}\right)_s \qquad (8.6)$$

since $\frac{1}{\rho} = v$, where v is the specific volume. With isentropic compressibility being

$K_s = -\frac{1}{v}\left(\frac{\partial v}{\partial p}\right)_s$, we obtain

$$a^2 = \frac{v}{K_s} = \frac{1}{\rho K_s}$$

or

$$K_s = \frac{1}{\rho a^2} \tag{8.7}$$

The isentropic compressibility is the inverse of ρa^2. Since ρ and a can readily be measured, Eq. 8.7 can be used to give the isentropic compressibility K_s.

8.2 SPECIFIC HEATS

8.2.1 SPECIFIC HEATS AT CONSTANT PRESSURE c_p AND CONSTANT VOLUME c_v

The specific heats c_p and c_v were defined as

$$c_p = \left(\frac{\partial h}{\partial T} \right)_p \quad c_v = \left(\frac{\partial u}{\partial T} \right)_v$$

The equations for entropy changes $ds(T,v)$ and $ds(T,\mathrm{p})$ given by Eqs. 7.52 and 7.53 in Chapter 7 and reproduced below are

$$ds(T,v) = c_v \left(\frac{dT}{T} \right) + \left(\frac{\partial p}{\partial T} \right)_v dv$$

$$ds(T,\mathrm{p}) = c_p \frac{dT}{T} - \left(\frac{\partial v}{\partial T} \right)_p dp$$

Equating them, we obtain

$$dT = \frac{T \left(\dfrac{\partial v}{\partial T} \right)_p}{c_p - c_v} dp + \frac{T \left(\dfrac{\partial p}{\partial T} \right)_v}{c_p - c_v} dv$$

The above can be written in terms of coefficient of volume expansion β and pressure coefficient α as

$$dT = \frac{T\beta v}{c_p - c_v} dp + \frac{\alpha p T}{c_p - c_v} dv \tag{8.8}$$

In the above equation, T is a function of p and v, namely, $T(p,v)$. Hence, $dT(p, v)$ is written as

$$dT(p,v) = \left(\frac{\partial T}{\partial p} \right)_v dp + \left(\frac{\partial T}{\partial v} \right)_p dv = \frac{dp}{\alpha p} + \frac{dv}{\beta v} \tag{8.9}$$

and further equating the coefficients of dp and dv in the above relationships given in Eqs. 8.8 and 8.9, we obtain

$$c_p - c_v = \beta v \alpha p T \qquad (8.10)$$

Thus, the difference in specific heats can be related to the $p - v - T$ data. Equation 8.10 expresses the difference in specific heats in terms of the thermodynamic coefficients α and β.

As an example, for an ideal gas for which the equation of state is $pv = RT$, we can obtain $\alpha = \dfrac{1}{p}\left(\dfrac{\partial p}{\partial T}\right)_v = \dfrac{1}{T}$ and $\beta = \dfrac{1}{v}\left(\dfrac{\partial v}{\partial T}\right)_p = \dfrac{1}{T}$ and Eq. 8.10 gives for an ideal gas

$$c_p - c_v = R \qquad (8.11)$$

which is a familiar relation between c_p and c_v for an ideal gas. If we define the specific heat capacities as per mole of gas and denote them by $\tilde{c}_p$ and $\tilde{c}_v$ and with the equation of state for an ideal gas being $p\tilde{v} = R_0 T$ where $\tilde{v}$ is the volume per mole of the gas, we get

$$\tilde{c}_p - \tilde{c}_v = R_0 \qquad (8.12)$$

From the definitions of α and β (i.e., Eqs. 8.1 and 8.4), we can write Eq. 8.10 in terms of partial derivatives as

$$c_p - c_v = T\left(\frac{\partial v}{\partial T}\right)_p \left(\frac{\partial p}{\partial T}\right)_v \qquad (8.13)$$

Using the cyclic rule for variables p, v and T we write

$$\left(\frac{\partial p}{\partial T}\right)_v \left(\frac{\partial T}{\partial v}\right)_p \left(\frac{\partial v}{\partial p}\right)_T = -1$$

$$\text{or} \left(\frac{\partial p}{\partial T}\right)_v = -\frac{1}{\left(\dfrac{\partial T}{\partial v}\right)_p \left(\dfrac{\partial v}{\partial p}\right)_T} = -\left(\frac{\partial v}{\partial T}\right)_p \left(\frac{\partial p}{\partial v}\right)_T$$

Hence, Eq. 8.13 can therefore be written as

$$c_p - c_v = -T\left(\frac{\partial v}{\partial T}\right)_p^2 \left(\frac{\partial p}{\partial v}\right)_T = vT\frac{\beta^2}{K_T} \qquad (8.14)$$

Since $\beta^2 = \dfrac{1}{v^2}\left(\dfrac{\partial v}{\partial T}\right)_p^2$ is always positive and the isothermal compressibility

$\kappa_T = -\dfrac{1}{v}\left(\dfrac{\partial v}{\partial p}\right)_T$ for all known substances is positive, Eq. 8.14 indicates that

$$c_p - c_v > 0$$

$$\text{or } c_p > c_v$$

for T positive. It is also seen from Eq. 8.14 that as the temperature tends to zero $(T \to 0)$ either $c_p \to c_v$ or both c_p and c_v become zero. In fact the specific heat tending to 0 as $T \to 0$ is one of the statements of the third law of thermodynamics.

Equation 8.14 also indicates that for incompressible substances c_p and c_v are identical. Note that for the particular case of water at 4°C, where the density of water is a maximum, that is, $\left(\dfrac{\partial v}{\partial T}\right)_p = 0$ and thus $c_p = c_v$ at 4°C.

Equation 8.14 is useful in that it permits c_p or c_v to be determined when one of the heat capacities is known. Since c_v, in general, is difficult to measure, Eq. 8.14 can be used to determine c_v when c_p is known.

8.2.2 RATIO OF SPECIFIC HEATS

Equations 7.52 and 7.53 in Chapter 7 gave the incremental change of entropy for variables T,v and T,p as

$$ds(T,v) = c_v \frac{dT}{T} + \left(\frac{\partial p}{\partial T}\right)_v dv$$

and

$$ds(T,p) = c_p \frac{dT}{T} - \left(\frac{\partial v}{\partial T}\right)_p dp$$

We use these two equations to obtain the ratio of specific heats for an isentropic process for which $ds = 0$, that is,

$$\frac{c_p}{c_v} = \frac{-\left(\dfrac{\partial v}{\partial T}\right)_p}{\left(\dfrac{\partial p}{\partial T}\right)_v} \frac{dp_s}{dv_s} = -\frac{\left(\dfrac{\partial v}{\partial T}\right)_p}{\left(\dfrac{\partial p}{\partial T}\right)_v}\left(\frac{\partial p}{\partial v}\right)_s$$

where the subscript "s" denotes constant entropy. But the cyclic rule for three variables p,v,T from which Eq. 8.5 was derived gives

$$\frac{\left(\dfrac{\partial v}{\partial T}\right)_p}{\left(\dfrac{\partial p}{\partial T}\right)_v} = -\left(\frac{\partial v}{\partial p}\right)_T = vk_T$$

Further since

$$k_s = -\frac{1}{v}\left(\frac{\partial v}{\partial p}\right)_s$$

we obtain for an isentropic process

$$\frac{c_p}{c_v} = \frac{k_T}{k_S} \tag{8.15}$$

The ratio of specific heats is equal to the ratio of the isothermal and isentropic compressibility k_T and k_s. For an ideal gas where the equation of state given by $pv = RT$, $k_T = \dfrac{1}{p}$ and $k_s = \dfrac{1}{\gamma p}$ yields

$$\frac{c_p}{c_v} = \gamma \tag{8.16}$$

the familiar relationship for the ratio of the specific heats for an ideal gas with constant c_p and c_v

8.2.3 VARIATION OF SPECIFIC HEATS c_v AND c_p WITH SPECIFIC VOLUME v AND PRESSURE p

We had shown in the last chapter that the internal energy per unit mass u could be expressed in terms of variables temperature T and specific volume v, that is $u(T,v)$. Since

$$c_v = \left(\frac{\partial u}{\partial T} \right)_v$$

c_v would also be a function of T and v, that is, $c_v(T,v)$. Writing

$$c_v = \left(\frac{\partial u}{\partial T} \right)_v = \left(\frac{\partial u}{\partial s} \right)_v \left(\frac{\partial s}{\partial T} \right)_v = T \left(\frac{\partial s}{\partial T} \right)_v$$

Differentiating the above with respect to v gives

$$\left(\frac{\partial c_v}{\partial v} \right)_T = \frac{\partial}{\partial v} \left(T \left(\frac{\partial s}{\partial T} \right)_v \right)_T = T \frac{\partial^2 s}{\partial v \partial T} = T \frac{\partial}{\partial T} \left(\left(\frac{\partial s}{\partial v} \right)_T \right)_v$$

and using the Maxwell relation $\left(\dfrac{\partial s}{\partial v} \right)_T = \left(\dfrac{\partial p}{\partial T} \right)_v$, the above becomes

$$\left(\frac{\partial c_v}{\partial v} \right)_T = T \frac{\partial}{\partial T} \left(\frac{\partial p}{\partial T} \right)_v$$

$$= T \left(\frac{\partial^2 p}{\partial T^2} \right)_v \tag{8.17}$$

Similarly for specific heat at constant pressure c_p, we have

$$c_p = \left(\frac{\partial h}{\partial T}\right)_p = \left(\frac{\partial u}{\partial s}\right)_p \left(\frac{\partial s}{\partial T}\right)_p = T\left(\frac{\partial s}{\partial T}\right)_p$$

$$\left(\frac{\partial c_p}{\partial p}\right)_T = \frac{\partial}{\partial p}\left(T\left(\frac{\partial s}{\partial T}\right)_p\right)_T = T\frac{\partial^2 s}{\partial p \partial T} = T\frac{\partial}{\partial T}\left(\left(\frac{\partial s}{\partial p}\right)_T\right)_p$$

and with the Maxwell relation $\left(\dfrac{\partial s}{\partial p}\right)_T = \left(\dfrac{\partial v}{\partial T}\right)_p$, the above becomes

$$\left(\frac{\partial c_p}{\partial p}\right)_T = T\frac{\partial}{\partial T}\left(\frac{\partial v}{\partial T}\right)_p$$

$$= T\left(\frac{\partial^2 v}{\partial T^2}\right)_p \tag{8.18}$$

The partial derivatives of c_v and c_p with respect to v and p are equal to $T\left(\dfrac{\partial^2 p}{\partial T^2}\right)_v$ and

$T\left(\dfrac{\partial^2 v}{\partial T^2}\right)_p$ respectively.

For an ideal gas where $pv = RT$, $\left(\dfrac{\partial^2 p}{\partial T^2}\right)_v = 0$ and $\left(\dfrac{\partial^2 v}{\partial T^2}\right)_p = 0$. Thus, c_v is not a function of v and is a function of only T, namely, $c_v(T)$ for an ideal gas. Similarly, c_p is not a function of p and is only a function of T, namely, $c_p(T)$ for an ideal gas. This is in accord with the previous result that for an ideal gas $u = u(T)$ and $h = h(T)$.

8.3   JOULE THOMSON COEFFICIENT

The Joule Thomson coefficient μ is defined as the change in temperature with decrease in pressure due to throttling

$$\mu = \left(\frac{\partial T}{\partial p}\right)_H \tag{8.19}$$

and can be measured in an adiabatic throttling experiment where a gas at high pressure is throttled down to a low pressure via a porous plug in an insulated pipe. The adiabatic throttling process is a constant enthalpy process. The Joule Thomson coefficient is related to the other thermodynamic coefficients and heat capacity in the following.

Considering the variables T, p and h, the cyclic rule gives

$$\left(\frac{\partial T}{\partial p}\right)_h \left(\frac{\partial p}{\partial h}\right)_T \left(\frac{\partial h}{\partial T}\right)_p = \mu c_p \left(\frac{\partial p}{\partial h}\right)_T = -1$$

Thus,

$$\mu = \frac{-1}{c_p \left(\dfrac{\partial p}{\partial T}\right)_T}$$

From the Gibbs function

$$g = h - Ts$$

we obtain

$$\left(\frac{\partial h}{\partial p}\right)_T = \left(\frac{\partial g}{\partial p}\right)_T + T\left(\frac{\partial s}{\partial p}\right)_T$$

and using the relationship $\left(\dfrac{\partial g}{\partial p}\right)_T = v$ and Maxwell's equation $\left(\dfrac{\partial s}{\partial p}\right)_T = -\left(\dfrac{\partial v}{\partial T}\right)_p$,

$$\left(\frac{\partial h}{\partial p}\right)_T = v - T\left(\frac{\partial v}{\partial T}\right)_p$$

Hence,

$$\mu = \frac{-1}{c_p \left(\dfrac{\partial p}{\partial h}\right)_T} = \frac{\left[T\left(\dfrac{\partial v}{\partial T}\right)_p - v\right]}{c_p} \tag{8.20}$$

With the volume coefficient being $\beta = \dfrac{1}{v}\left(\dfrac{\partial v}{\partial T}\right)_p$, the Joule Thomson coefficient

becomes

$$\mu = \frac{v(\beta T - 1)}{c_p} \tag{8.21}$$

Knowledge of c_p and β permit the Joule Thomson coefficient to be found. It is seen from Eq. 8.20 that for an ideal gas where $\left(\dfrac{\partial v}{\partial T}\right)_p = \dfrac{R}{p}$

$$\mu = \left(\frac{RT}{p} - v\right)\frac{1}{c_p} = 0 \tag{8.22}$$

Thus, $\mu = 0$ for an ideal gas and there is no change in temperature for a throttling process of an ideal gas. However, for a real gas, $\mu \neq 0$ and μ can be positive or negative depending on the initial condition upstream of throttling.

The constant enthalpy curves in a plot of temperature and pressure corresponding to different downstream and upstream pressures are illustrated in Figure 8.1. For a

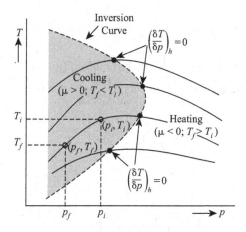

FIGURE 8.1   Iso-enthalpy curves and Joule Thomson expansion.

given upstream pressure p_i and temperature T_i, the downstream or final temperature T_f (after throttling) either increases or decreases depending on the upstream and downstream values of pressures. For higher values of p_i, the downstream temperature T_f initially increases as p_f is reduced. At some value of the downstream pressure p_f, the downstream temperature T_f reaches a maximum and subsequent reduction in pressure p_f leads to a decrease of temperature. This is observed in Figure 8.1 for all iso-enthalpy curves. The temperature at which the Joule Thomson coefficient changes sign is known as inversion temperature and from Eq. 8.21, it is equal to $T = \dfrac{1}{\beta}$ corresponding to $\mu = 0$. The locus of the maximum temperature values for the different values of enthalpy, shown dotted in Figure (8.1), is referred to as the inversion curve.

In the left side of the inversion curve wherein the temperature decreases (shown in gray in Figure 8.1), a reduction of pressure leads to cooling. When the temperature of the fluid, while undergoing iso-enthalpic expansion, is above the inversion temperature, the coefficient is positive and the temperature of the expanded fluid will increase. It is necessary to lower the temperature below the inversion temperature while using the Joule Thomson expansion process for liquefaction of gases.

8.4   THERMODYNAMIC COEFFICIENTS FOR DENSE GASES

We had discussed the thermodynamic functions and properties for real gases in the last chapter. In a similar way, we discuss the influence of non-ideality for the thermodynamic coefficients. Let us consider a dense gas for which the van der Waal's equation of state given by

$$\left(p + \frac{a}{\tilde{v}^2}\right)(\tilde{v} - b) = R_0 T \tag{8.23}$$

is valid. We obtained from Eq. 8.23,

$$p = \frac{R_0 T}{\tilde{v} - b} - \frac{a}{\tilde{v}^2} \text{ and } \left(\frac{\partial p}{\partial T} \right)_{\tilde{v}} = \frac{R_0}{\tilde{v} - b}$$

The coefficient of volume expansion can be obtained from the above; but first we use the cyclic rule for variables $\tilde{v}, T, p$ to obtain $\left(\frac{\partial \tilde{v}}{\partial T} \right)_p$, that is,

$$\left(\frac{\partial \tilde{v}}{\partial T} \right)_p \left(\frac{\partial T}{\partial p} \right)_{\tilde{v}} \left(\frac{\partial p}{\partial \tilde{v}} \right)_T = -1$$

$$\left(\frac{\partial \tilde{v}}{\partial T} \right)_p = \frac{-1}{\left(\frac{\partial T}{\partial p} \right)_{\tilde{v}} \left(\frac{\partial p}{\partial \tilde{v}} \right)_T} = -\frac{\left(\frac{\partial p}{\partial T} \right)_{\tilde{v}}}{\left(\frac{\partial p}{\partial \tilde{v}} \right)_T} \tag{8.24}$$

From the van der Waals equation $p = \frac{R_0 T}{\tilde{v} - b} - \frac{a}{\tilde{v}^2}$, we obtain

$$\left(\frac{\partial p}{\partial \tilde{v}} \right)_T = -\frac{R_0 T}{(\tilde{v} - b)^2} + \frac{2a}{\tilde{v}^3} = \frac{2a(\tilde{v} - b)^2 - R_0 T \tilde{v}^3}{\tilde{v}^3 (\tilde{v} - b)^2} \tag{8.25}$$

Using the above and $\left(\frac{\partial p}{\partial T} \right)_{\tilde{v}} = \frac{R_0}{\tilde{v} - b}$, we then write

$$\left(\frac{\partial \tilde{v}}{\partial T} \right)_p = -\frac{\left(\frac{\partial p}{\partial T} \right)_{\tilde{v}}}{\left(\frac{\partial p}{\partial \tilde{v}} \right)_T} = -\frac{\dfrac{R_0}{(\tilde{v} - b)}}{\dfrac{2a(\tilde{v} - b)^2 - R_0 T \tilde{v}^3}{\tilde{v}^3 (\tilde{v} - b)^2}} = \frac{R_0 \tilde{v}^3 (\tilde{v} - b)}{R_0 T \tilde{v}^3 - 2a(\tilde{v} - b)^2} \tag{8.26}$$

We thus obtain the pressure coefficient $\alpha = \frac{1}{p} \left(\frac{\partial p}{\partial T} \right)_v$, the coefficient of volume expansion $\beta = \frac{1}{v} \left(\frac{\partial v}{\partial T} \right)_p$ and isothermal compressibility $k_T = -\frac{1}{v} \left(\frac{\partial v}{\partial p} \right)_T$

The Joule Thomson coefficient, that is,

$$\mu = \frac{\left[T \left(\frac{\partial v}{\partial T} \right)_p - v \right]}{c_p}$$

can also be obtained from Eq. 8.26 as

$$\mu = \frac{1}{\tilde{c}_p} \left[\frac{2a\tilde{v}(\tilde{v} - b)^2 R_0 T b \tilde{v}^3}{R_0 T \tilde{v}^3 - 2a(\tilde{v} - b)^2} \right] \text{.} \tag{8.27}$$

It may be noted that in the above equation, the specific heat is on a per unit mole basis instead of per unit mass and the universal gas constant R_0 is used instead of the specific gas constant R.

9 Thermodynamic Equilibrium

9.1 INTRODUCTION

The state of a system is defined by a set of thermodynamic variables. For a simple compressible system, the state is defined by the internal energy, the volume and the number of moles n of the substance in the system. Subsequent to an interaction with the environment (e.g., heat and work exchange), the state of a system evolves from an initial state to a final equilibrium state.

There are various kinds of equilibrium, for example, thermal, mechanical, chemical, phase, etc. The equilibration times are different for the different kinds of equilibrium and the equilibration times can differ by orders of magnitude. For example, a mixture of H_2 and O_2 can be at thermal and mechanical equilibrium with its environment; however, chemical equilibrium of the reactive mixture at room temperature takes an "infinite" time to achieve. The composition of the system remains "constant" for an indefinite period. But, if a platinized gauze is placed in the system, reactions proceed rapidly to a final equilibrium mixture of H_2, O_2 and H_2O. Thus, the systems are often in partial equilibrium. However, analysis can be carried out with the thermodynamic variables of the "non-equilibrium" system as if the system is in equilibrium.

Thermodynamic equilibrium is defined when all the various kinds of equilibrium (i.e., thermal, mechanical, chemical) are attained.

9.2 EQUILIBRIUM CRITERION

A fundamental problem in thermodynamics is to determine the final equilibrium state subsequent to a process. Perhaps the simplest criterion to determine the final equilibrium state is via an extremum principle, that is, the thermodynamic variables are those that maximize (or minimize) some thermodynamic function. The most important is perhaps the maximum entropy criterion for an isolated system. The second law states that for an isolated system, spontaneous processes tend to increase the entropy function "S". Thus,

$$\Delta S \geq 0 \tag{9.1}$$

provides a direction for the evolution of an isolated system and at equilibrium, S = maximum. Since the entropy is an extremum at equilibrium, small departure from equilibrium gives

$$dS = 0 \tag{9.2}$$

DOI: 10.1201/9781003224044-9

The entropy of a composite system is additive over the constituent subsystems, that is,

$$S = \sum_i S_i$$

where S_i is the entropy of the ith subsystem. The additive property of the subsystems requires that entropy of a simple system to be first-order homogeneous function of the extensive parameters. That is if the extensive parameters are scaled by a factor λ, the entropy is scaled by the same factor λ (Euler's theorem), that is,

$$S(\lambda U, \lambda V, \lambda n) = \lambda S(U, V, n)$$

The entropy is continuous and differentiable and is a monotonically increasing function of energy. This implies that the partial derivative $\left(\dfrac{\partial S}{\partial U}\right)_{v,n}$ is a positive quantity

$$\left(\frac{\partial S}{\partial U}\right)_{v,n} > 0$$

The above derivative is the reciprocal of temperature, thus the above implies that the temperature is always positive. The entropy vanishes at the state when temperature approaches zero in accordance with the Nernst postulate (third law of thermodynamics).

From the second law, we may derive the equilibrium criteria for systems other than isolated system. For example, consider a simple system in contact with a heat reservoir at temperature T. Taking the system together with the reservoir as an isolated system, then the second law can be written as

$$\Delta S^0 = (\Delta S + \Delta S_R) \geq 0$$

where S^0, S and S_R are the entropies of the combined system, entropy of the system and reservoir, respectively.

If a process receives heat ΔQ from the reservoir, the entropy change of the reservoir is

$$\Delta S_R = -\frac{\Delta Q}{T},$$

and from the second law

$$\Delta S + \Delta S_R = \Delta S - \frac{\Delta Q}{T} \geq 0$$

$$\Delta Q \leq T\Delta S$$

The first law for the system can be written as

$$\Delta U = \Delta Q - W$$

and hence,

$$W \leq -\Delta U + T\Delta S \leq -\Delta(U - TS) - S\Delta T$$

where W is the work done by the system. If we separate the work as expansion work $p\Delta V$ and other kinds of work W' (e.g., electromagnetic, paddle work, surface tension), then the above expression gives

$$W' \leq -\Delta A - S\Delta T - p\Delta V \tag{9.3}$$

where $A = U - TS$ is defined as the Helmholtz function. For an isothermal, isochoric process $\Delta T = 0$ and $\Delta V = 0$ we write

$$W' \leq -\Delta A \tag{9.4}$$

that is, the non-expansion work W' is less than or equalled to the decrease of the Helmholtz function. The equality sign applies when the process is reversible. If there is no other form of work done by the system, that is, $W' = 0$, and Eq. 9.4 gives

$$0 \leq -\Delta A \tag{9.5}$$

The above gives the direction of evolution of the state of the system towards equilibrium and at equilibrium, $A = $ minimum.

For a system interacting with both a heat and a large pressure reservoir, Eq. 9.3 gives

$$W' \leq -\Delta A - S\Delta T - \Delta(pV) + V\Delta p$$

and rearranging the above for an isothermal process gives

$$W' \leq -\Delta(U - TS + pV) + V\Delta p$$

Defining the Gibbs function (or the Gibbs-free energy) as

$$G = U - TS + pV$$

the above becomes

$$W' \leq -\Delta G + V\Delta p$$

For an isobaric process $p = $ constant, $\Delta p = 0$; thus

$$W' \leq -\Delta G$$

If there is no other kind of work other than the $p\Delta V$ expansion work, then

$$0 \leq -\Delta G \tag{9.6}$$

Thus for a system interacting with a heat and pressure reservoir, the evolution of the system toward equilibrium is in the direction of decreasing Gibbs-free energy and at equilibrium G = minimum.

Summarizing, we see that for an isolated system, approach to equilibrium is toward the increase in the entropy and S = maximum at equilibrium. For a system interacting with a single heat reservoir, the approach to equilibrium for a system undergoing an isothermal, isochoric process is via the decrease of the Helmholtz function and at equilibrium, A = minimum. Similarly, a system undergoing an isothermal, isobaric process, the approach to equilibrium is via the decrease of the Gibbs-free energy. At equilibrium, the Gibbs-free energy G = minimum.

9.3   THERMAL EQUILIBRIUM

Consider an isolated system $(A + B)$ consisting of two sub-systems A and B separated by a partition as shown in Figure 9.1. Let the partition between A and B be rigid and impermeable to mass exchange. Thus, the volumes V_A and V_B are constant, and the number of moles of the gas in each of the volumes n_A and n_B are also constant. Let the partition be diathermal so that heat exchange can take place; then the net internal energy $U = U_A + U_B$ = constant, but U_A and U_B are not constant since A and B can exchange energy.

According to the second law, the evolution of the isolated system toward equilibrium is

$$\Delta S = \Delta S_A + \Delta S_B \geq 0 \qquad (9.7)$$

Since for A, $S_A(U_A, V_A, n_A)$ and for B, $S_B(U_B, V_B, n_B)$, we write the net incremental change in entropy as

$$dS = dS_A + dS_B$$

$$= \left(\frac{\partial S_A}{\partial U_A}\right) dU_A + \left(\frac{\partial S_A}{\partial V_A}\right) dV_A + \left(\frac{\partial S_A}{\partial n_A}\right) dn_A + \left(\frac{\partial S_B}{\partial U_B}\right) dU_B + \left(\frac{\partial S_B}{\partial V_B}\right) dV_B + \left(\frac{\partial S_B}{\partial n_B}\right) dn_B \qquad (9.8)$$

Since $dV_A = dV_B = 0$, $dn_{Ai} = dn_{Bi} = 0$ and $dU_A = -dU_B$ $(U_A + U_B$ = constant$)$, we can write

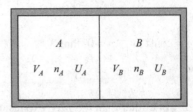

FIGURE 9.1   Isolated system with two subsystems separated by a diathermal partition.

$$dS = \left(\frac{\partial S_A}{\partial U_A} - \frac{\partial S_B}{\partial U_B} \right) dU_A$$

and at equilibrium where $dS = 0$, we obtained

$$\left(\frac{\partial S_A}{\partial U_A} - \frac{\partial S_B}{\partial U_B} \right) = 0 \tag{9.9}$$

since dU_A is arbitrary. For two systems in thermal contact, experience indicates that the temperature of the two systems is equalled. Thus, the derivative $\left(\frac{\partial S}{\partial U} \right)$ must be some function of temperature.

We define

$$\left(\frac{\partial S}{\partial U} \right) = \frac{1}{T} \tag{9.10}$$

and at equilibrium,

$$\left(\frac{1}{T_A} - \frac{1}{T_B} \right) = 0$$

or $T_A = T_B$. According to the second law $\Delta S \geq 0$, we obtained

$$\left[\left(\frac{\partial S_A}{\partial U_A} \right) - \left(\frac{\partial S_B}{\partial U_B} \right) \right] \Delta U_A \geq 0$$

$$\text{or} \left(\frac{1}{T_A} - \frac{1}{T_B} \right) dU_A \geq 0 \tag{9.11}$$

Thus if $T_A > T_B$, $\frac{1}{T_A} < \frac{1}{T_B}$, hence $dU_A < 0$. This agrees with experience that if system A is hotter than system B, heat is transferred from A to B and the internal energy of A decreases.

If "S" is a maximum at equilibrium, the derivative $\left(\frac{\partial S}{\partial U} \right) = 0$ and the second derivative $\left(\frac{\partial^2 S}{\partial U^2} \right) < 0$. Taking the second derivative of $\left(\frac{\partial S}{\partial U} \right)$, we obtained

$$\left(\frac{\partial^2 S_A}{\partial U_A} \right) dU^2{}_A + \left(\frac{\partial^2 S_B}{\partial U^2{}_B} \right) dU^2{}_B$$

If A and B are similar systems, the necessary condition for a maximum requires that

$$\frac{\partial^2 S}{\partial U^2} < 0$$

for both A and B.

Since $\left(\dfrac{\partial S}{\partial U}\right) = \dfrac{1}{T}$, the second derivative gives

$$\frac{\partial^2 S}{\partial U^2} = \frac{\partial\left(\frac{1}{T}\right)}{\partial U} = -\frac{1}{T^2}\left(\frac{\partial T}{\partial U}\right)$$

As T is positive, we see that for $\left(\dfrac{\partial^2 S}{\partial U^2}\right) < 0$, $\left(\dfrac{\partial T}{\partial U}\right) > 0$. The internal energy U is an increasing function of temperature, for example, for a perfect gas $U = C_V T$. Thus, $\left(\dfrac{\partial T}{\partial U}\right) = \dfrac{1}{C_V}$ and since $C_v > 0$, $\left(\dfrac{\partial^2 S}{\partial U^2}\right) < 0$. Hence, the entropy of systems A and B is indeed a maximum at equilibrium.

9.4   MECHANICAL EQUILIBRIUM

If the partition separating A and B is movable, then V_A and V_B are no longer constant. But $V' = V_A + V_B = $ constant, we have $dV_A = -dV_B$. If the partition is diathermal and energy can exchange between A and B, we have $dU_A = -dU_B$. Thus, dS is

$$dS = dS_A + dS_B$$

$$= \left(\frac{\partial S_A}{\partial U_A}\right)dU_A + \left(\frac{\partial S_A}{\partial V_A}\right)dV_A + \left(\frac{\partial S_A}{\partial n_A}\right)dn_A + \left(\frac{\partial S_B}{\partial U_B}\right)dU_B + \left(\frac{\partial S_B}{\partial V_B}\right)dV_B + \left(\frac{\partial S_B}{\partial n_B}\right)dn_B$$

$$\text{(9.12)}$$

and with, $dU_A = -dU_B$, $dV_A = -dV_B$ and $dn_A = dn_B = 0$, the above becomes

$$dS = \left[\left(\frac{\partial S_A}{\partial U_A}\right) - \left(\frac{\partial S_B}{\partial U_B}\right)\right]dU_A + \left[\left(\frac{\partial S_A}{\partial V_A}\right) - \left(\frac{\partial S_B}{\partial V_B}\right)\right]dV_A$$

At equilibrium where $dS = 0$ and dU_A, dV_A, are arbitrary

$$\left[\left(\frac{\partial S_A}{\partial U_A}\right) - \left(\frac{\partial S_B}{\partial U_B}\right)\right] = 0 \qquad\qquad\qquad\qquad \text{(9.13)}$$

$$\left[\left(\frac{\partial S_A}{\partial V_A}\right) - \left(\frac{\partial S_B}{\partial V_B}\right)\right] = 0 \qquad\qquad\qquad\qquad \text{(9.14)}$$

As described previously, for thermal equilibrium, $T_A = T_B$, and thus for mechanical equilibrium, $\left[\left(\dfrac{\partial S_A}{\partial V_A}\right) - \left(\dfrac{\partial S_B}{\partial V_B}\right)\right] = 0$. Experience indicates that the pressure across the partition is equaled, thus $\left(\dfrac{\partial S}{\partial V}\right)$ must be some function of the pressure "p". From dimensional consideration, the units of $\left(\dfrac{\partial S}{\partial V}\right)$ is pressure/temperature. Thus, we define

$$\left(\frac{\partial S}{\partial V}\right) = \frac{p}{T} \tag{9.15}$$

and mechanical equilibrium gives

$$\left(\frac{p_A}{T_A} - \frac{p_B}{T_B}\right) = 0 \tag{9.16}$$

Since the partition is diathermal and $T_A = T_B$, we see that at equilibrium, $p_A = p_B$ in accord with experiments. With $\Delta S > 0$, we have

$$\left[\left(\frac{\partial S_A}{\partial V_A}\right) - \left(\frac{\partial S_B}{\partial V_B}\right)\right] dV_A > 0$$

or

$$\left(\frac{p_A}{T_A} - \frac{p_B}{T_B}\right) dV_A > 0 \tag{9.17}$$

Thus if $p_A > p_B$, $\left(\dfrac{p_A}{T_A} - \dfrac{p_B}{T_B}\right) > 0$, since $T_A = T_B$. Thus, $dV_A > 0$, that is, if the pressure of A is greater than B, then the volume of A increases in accordance with experiments.

9.5   EQUILIBRIUM WITH MASS EXCHANGE

If the rigid partition separating A and B (Figure 9.1) is permeable to mass exchange, then n_A and n_B will not be constant, but since $n = n_A + n_B = $ constant, $dn_A = -dn_B$. Thus for a rigid, diathermal and permeable partition, we have $dU_A = -dU_B$, $dV_A = -dV_B = 0$ and $dn_A = -dn_B$. At equilibrium where $dS = 0$,

$$dS = \left(\frac{1}{T_A} - \frac{1}{T_B}\right) dU_A + \left[\left(\frac{\partial S_A}{\partial n_A}\right) - \left(\frac{\partial S_B}{\partial n_B}\right)\right] dn_A = 0 \tag{9.18}$$

Since $T_A = T_B$, at equilibrium,

$$\left[\left(\frac{\partial S_A}{\partial n_A}\right) - \left(\frac{\partial S_B}{\partial n_B}\right)\right] = 0 \tag{9.19}$$

as dn_A is arbitrary. We define

$$\left(\frac{\partial S}{\partial n}\right) = -\left(\frac{\mu}{T}\right) \tag{9.20}$$

where μ is known as chemical potential. Thus,

$$\left(\frac{\mu_B}{T_B} - \frac{\mu_A}{T_A}\right) dn_A = 0 \tag{9.21}$$

and $T_A = T_B$, we have $\mu_A = \mu_B$ at equilibrium.

According to the second law wherein $\Delta S > 0$ for the composite isolated system, we obtained

$$\left(\frac{\mu_B}{T_B} - \frac{\mu_A}{T_A}\right) dn_A > 0 \tag{9.22}$$

Thus if $\mu_A > \mu_B$, the bracketed term < 0 and $dn_A < 0$. In other words, if the chemical potential μ_A is greater than μ_B, $dn_A < 0$ and the mass flows from A to B in accordance with thermal equilibrium, wherein energy is transferred from a higher to a lower temperature $(T_A > T_B, dU_A < 0)$.

9.6 CHEMICAL POTENTIAL

In the previous section, we have considered the equilibrium of a system with mass exchange and introduced the term chemical potential $\mu = -T\left(\dfrac{\partial S}{\partial n}\right)_{V,U}$. The chemical potential is important in the discussion of equilibrium of systems having multiple component and phases. We wish to generalize the previous discussion to a more general system of multiple components and multiple phases. A phase is defined as a homogeneous and distinct part of a system, which is separated from the other part by interfaces. Thus in the example considered in the previous section, subsystems A and B can be thought as two phases separated by an interface across which particle or mass exchange takes place.

As discussed in the previous chapters, the combination of the first and second law for a closed system (of constant mass) is written as

$$dU = TdS - pdV.$$

Hence $U(S,V)$ and we can write

$$dU = \left(\frac{\partial U}{\partial S}\right)_V dS + \left(\frac{\partial U}{\partial V}\right)_S dV$$

giving

$$T = \left(\frac{\partial U}{\partial S}\right)_V, p = -\left(\frac{\partial U}{\partial V}\right)_S. \tag{9.23}$$

Alternatively, we can write the fundamental equation as $S(U,V)$ and obtained

$$dS = \frac{1}{T}dU + \frac{p}{T}dV$$

and

$$dS = \left(\frac{\partial S}{\partial U}\right)_V dU + \left(\frac{\partial S}{\partial V}\right)_U dV$$

giving

$$\frac{1}{T} = \left(\frac{\partial S}{\partial U}\right)_V , \frac{p}{T} = \left(\frac{\partial S}{\partial V}\right)_U \qquad (9.24)$$

Of importance is the enthalpy function $H = U + pV$, where we can write

$$dH = TdS + Vdp$$

and

$$dH = \left(\frac{\partial H}{\partial S}\right)_p dS + \left(\frac{\partial H}{\partial p}\right)_S dp$$

giving

$$T = \left(\frac{\partial H}{\partial S}\right)_p , V = \left(\frac{\partial H}{\partial p}\right)_S \qquad (9.25)$$

The two other thermodynamic functions of importance are the Helmholtz function $A = U - TS$ and the Gibbs function (or free energy) $G = U + pV - TS = H - TS$. Combining the above with the first and second laws gives

$$dA = -SdT - pdV$$

$$= \left(\frac{\partial A}{\partial T}\right)_V dT + \left(\frac{\partial A}{\partial V}\right)_T dV$$

and equating the coefficients give

$$S = -\left(\frac{\partial A}{\partial T}\right)_V , p = -\left(\frac{\partial A}{\partial V}\right)_T \qquad (9.26)$$

Similarly for the Gibbs-free energy, we write

$$dG = -SdT + Vdp$$

$$= \left(\frac{\partial G}{\partial T}\right)_p dT + \left(\frac{\partial G}{\partial p}\right)_T dp$$

which gives

$$S = -\left(\frac{\partial G}{\partial T}\right)_p , V = \left(\frac{\partial G}{\partial p}\right)_T , \qquad (9.27)$$

We now consider a multicomponent system or phase where the number of moles of the substance can vary (e.g., mass diffusion, chemical reactions). The state functions are now dependent on the number of moles n_i of the various components in addition to U, V, T, p, etc. Thus we write $U(S, V, n_1, n_2, n_3, - - - n_i, -)$ and accordingly for component i

$$dU\ (S, V, n_i) = \left(\frac{\partial U}{\partial S}\right)_{V, n_i} dS + \left(\frac{\partial U}{\partial V}\right)_{S, n_i} dV + \sum_i \left(\frac{\partial U}{\partial n_i}\right)_{S, V, n_{j;j\neq i}} dn_i$$

$$= TdS - pdV + \sum_i \mu_i dn_i \qquad (9.28)$$

where

$$T = \left(\frac{\partial U}{\partial S}\right)_{V, n_i}, p = -\left(\frac{\partial U}{\partial V}\right)_{U, n_i}, \mu_i = \left(\frac{\partial U}{\partial n_i}\right)_{S, V, n_{j;j\neq i}} \qquad (9.29)$$

In Eq. 9.28 for the change of internal energy, $\mu_i dn_i$ can be thought of as work done on the system when dn_i moles of component "i" is added to the system.

Similar to Eq. 9.28, we may write

$$dS(U, V, n_i) = \left(\frac{\partial S}{\partial U}\right)_{V, n_i} dU + \left(\frac{\partial S}{\partial V}\right)_{U, n_i} dV + \sum_i \left(\frac{\partial S}{\partial n_i}\right)_{U, V, n_{j;j\neq i}} dn_i$$

$$= \frac{1}{T} dU + \frac{p}{T} dV - \sum_i \frac{\mu_i}{T} dn_i$$

Thus,

$$\frac{1}{T} = \left(\frac{\partial S}{\partial U}\right)_{V, n_i}, \frac{p}{T} = \left(\frac{\partial S}{\partial V}\right)_{U, n_i}, \mu_i = -T\left(\frac{\partial S}{\partial n_i}\right)_{U, V, n_{j;j\neq i}} \qquad (9.30)$$

The expressions for the Helmholtz and Gibbs functions can be similarly obtained as

$$dA(T, V, n_i) = \left(\frac{\partial A}{\partial T}\right)_{V, n_i} dT + \left(\frac{\partial A}{\partial V}\right)_{T, n_i} dV + \sum_i \left(\frac{\partial A}{\partial n_i}\right)_{T, V, n_{j;j\neq i}} dn_i$$

$$= -SdT - pdV + \sum_i \mu_i dn_i$$

giving

$$S = -\left(\frac{\partial A}{\partial T}\right)_{V, n_i}, p = -\left(\frac{\partial A}{\partial V}\right)_{T, n_i}, \mu_i = \left(\frac{\partial A}{\partial n_i}\right)_{T, V, n_{j;j\neq i}} \qquad (9.31)$$

$$dG(T,p,n_i) = \left(\frac{\partial G}{\partial T}\right)_{p,n_i} dT + \left(\frac{\partial G}{\partial p}\right)_{T,n_i} dp + \sum_i \left(\frac{\partial G}{\partial n_i}\right)_{T,p,n_{j;j\neq i}} dn_i$$

$$= -SdT + Vdp + \sum_i \mu_i dn_i$$

. giving

$$S = -\left(\frac{\partial G}{\partial T}\right)_{p,n_i} , V = \left(\frac{\partial G}{\partial p}\right)_{T,n_i} , \mu_i = \left(\frac{\partial G}{\partial n_i}\right)_{T,p,n_{j;j\neq i}} \tag{9.32}$$

Summarizing the chemical potential can be expressed in different variables depending on the thermodynamic potential involved, that is,

$$\mu_i = \left(\frac{\partial U}{\partial n_i}\right)_{S,V,n_{j;j\neq i}} = T\left(\frac{\partial S}{\partial n_i}\right)_{U,V,n_{j;j\neq i}} = \left(\frac{\partial A}{\partial n_i}\right)_{T,V,n_{j;j\neq i}} = \left(\frac{\partial G}{\partial n_i}\right)_{T,p,n_{j;j\neq i}} \tag{9.33}$$

If only one species $n_i = n$ is present, we write

$$\mu_i = \left(\frac{\partial G}{\partial n_i}\right)_{T,p} = \frac{G}{n_i} = \tilde{g}_i(T,p) \tag{9.34}$$

where $\tilde{g}_i(T,p)$ is the specific Gibbs function per mole.

Since the fundamental equations are homogeneous first-order equations, they have the property that the extensive variables are scaled by a constant λ, then the function itself is scaled by λ, for example as considered at the beginning of this chapter,

$$U(\lambda S, \lambda V, \lambda n_i) = \lambda U(S,V,n_i)$$

This is known as Euler's equation and to prove this we can differentiate the above equation by λ, that is,

$$\frac{\partial U(\lambda S, \lambda V, \lambda n_i)}{\partial(\lambda S)}\frac{\partial(\lambda S)}{\partial \lambda} + \frac{\partial U(\lambda S, \lambda V, \lambda n_i)}{\partial(\lambda V)}\frac{\partial(\lambda V)}{\partial \lambda} +$$

$$\sum \frac{\partial U(\lambda S, \lambda V, \lambda n_i)}{\partial(\lambda n_i)}\frac{\partial(\lambda n_i)}{\partial \lambda} = U(S,V,n_i) \tag{9.35}$$

Simplifying, we get

$$\frac{\partial U(\lambda S, \lambda V, \lambda n_i)}{\partial(\lambda S)}S + \frac{\partial U(\lambda S, \lambda V, \lambda n_i)}{\partial(\lambda V)}V + \sum \frac{\partial U(\lambda S, \lambda V, \lambda n_i)}{\partial(\lambda n_i)}n_i$$

$$= U(S,V,n_i) \tag{9.36}$$

Since the above holds for any value of λ, we take $\lambda = 1$ and obtained

$$\left(\frac{\partial U}{\partial S}\right)S + \left(\frac{\partial U}{\partial V}\right)V + \sum\left(\frac{\partial U}{\partial n_i}\right)n_i = U\ (S,V,n_i)$$

$$\text{or } TS - pV + \sum \mu_i n_i = U$$

Differentiating the above gives

$$TdS + SdT - pdV - Vdp + \sum \mu_i dn_i + \sum n_i d\mu_i = dU \qquad (9.37)$$

and noting that

$$dU = TdS - pdV + \sum \mu_i dn_i$$

the above equation reduces to

$$SdT - Vdp + \sum n_i d\mu_i = 0 \qquad (9.38)$$

The above is known as Gibbs Duhem equation. Similarly, since $G = U + pV - TS$, we have on substituting for $U = TS - pV + \sum \mu_i n_i$ in the expression for G

$$G = TS - pV + \sum \mu_i n_i + pV - TS$$

$$= \sum \mu_i n_i \qquad (9.39)$$

The above is also referred to as Gibbs Duhem equation. And for one component when $n_i = n$, and all other $n_i = 0$, we have

$$G = \mu n$$

and

$$\mu = \frac{G}{n} = \tilde{g}(T,p) \qquad (9.40)$$

where $\tilde{g}(T,p)$ is the specific Gibbs function per mole.

10 Equilibrium of Species in a Chemically Reacting System

10.1 INTRODUCTION

In chemically reacting systems, new species are formed and heat is either released or absorbed from the environment. The release or absorption of heat is dependent on the species formed in the reaction. The equilibrium concentration of the different species in the reaction is governed by the pressure and temperature of the system. The criterion for thermodynamic equilibrium for non-reacting systems, that is chemically inert systems of the previous chapter, is extended for chemical reacting systems and the concentration of the species is determined.

10.2 CHOICE OF BASIC DATUM FOR THE STATE FUNCTIONS AND HEAT OF FORMATION

In a non-reacting system, the concentration of the various species of the mixture do not change. The choice of a reference base for the state functions (i.e., enthalpy, internal energy) can be arbitrary since when calculating the difference of the state function between two equilibrium states, the reference datum cancels out. However, for chemically reacting systems, species are destroyed and new ones are formed. It becomes necessary to seek a common reference state for all substances. The standard reference state is 298 K and 1 atm. pressure.

In general, heat is generated when a compound is formed. It is necessary to take this energy release into consideration when defining the state function at the reference state. As an example, consider the formation of 1 mole of H_2O from the reaction $H_2 + \frac{1}{2} O_2 = H_2O$ at the reference state of 298 K and 1 atm. pressure. The enthalpy change for the reaction at the reference state can be written as

$$\Delta H = \tilde{h}_{H_2O}(298) - \left\{ \tilde{h}_{H_2}(298) + \frac{1}{2}\tilde{h}_{O_2}(298) \right\} \tag{10.1}$$

where $\tilde{h}$ is the enthalpy per mole of the particular species as indicated by the subscripts.

If we were to arbitrarily assign zero values of enthalpy for all species at the reference state of 298 K and 1 atm. pressure, then enthalpy or heat release $\Delta H = 0$. However, experiments indicate that 286.7 kJ are released in the reaction at the reference state. To account for the heat release when a substance is formed, the enthalpy of formation is used.

The enthalpy of formation of a substance is defined as the heat release when 1 mole of the substance is formed from its naturally occurring stable elements at the reference state (298 K and 1 atm.). The naturally occurring elements are gaseous oxygen O_2, nitrogen gas N_2, chlorine gas Cl_2, solid carbon C(s), etc. They are assigned zero values for the enthalpy of formation at the reference state.

The enthalpy of formation is denoted by $\Delta \tilde{h}_f{}^0$ where $\Delta \tilde{h}_f$ indicates the enthalpy per mole of the substance over and above of its naturally occurring elements required to form it, while the superscript "0" denotes that it is formed at the reference state. Thus, we write for the heat of formation at temperature T assuming ideal gas and constant specific heat to be valid as

$$\tilde{h}_f(T) = \Delta \tilde{h}_f{}^0 + \int_{298}^{T} \tilde{c}_p \, dT \tag{10.2}$$

For the case of H_2O, we write using Eq. 10.1 for the enthalpy change per mole of H_2O formed at the reference state as

$$\Delta \tilde{h}_{f,H2O}^0 = \tilde{h}_{f,H2O}^0(298) - [\tilde{h}_{f,H2}^0(298) + 1/2 \, \tilde{h}_{f,O2}^0(298)]$$

Since 286.7 kJ/mole is released in the reaction at 298 K, the enthalpy required to form one mole of H_2O from one mole of H_2 and ½ mole of O_2 is negative and therefore

$$\Delta \tilde{h}_{f,H2O}^0 = -286.7 \text{ kJ/mole-}$$

The heat of formation $\Delta \tilde{h}_f{}^0$ is at the reference state and is also known as the standard heat of formation.

Similar to the enthalpy of formation, the internal energy of formation is defined as the internal energy required to form one mole of the substance at the standard state from its naturally occurring elements at the same standard state. It is denoted by $\Delta \tilde{u}_f{}^0$.

The relation between the internal energy of formation and enthalpy of formation is

$$\Delta \tilde{u}_f{}^0 = \Delta \tilde{h}_f{}^0 - p\tilde{v}$$

where the pressure p is 1 atm. If the substance is an ideal gas,

$$\Delta \tilde{u}_f{}^0 = \Delta \tilde{h}_f^0 - R_0 T_0$$

with T_0 being 298 K.

The heat of formation and internal energy of formation are properties characterizing the chemical structure of the substance. The standard heat of formation is readily available in literature for different substances.

The superscript "~" in the heat of formation and internal energy of formation, which denotes per mole, is generally omitted and the standard heat and internal energy of formation per mole of the substance are usually denoted by Δh_f^0 and Δu_f^0, respectively.

10.3   ENTROPY OF THE SPECIES IN A CHEMICAL REACTION: THIRD LAW OF THERMODYNAMICS

The third law of thermodynamics is used to define the absolute values of entropy. The third law states that all pure crystalline substances have zero value for the entropy at the absolute zero temperature. At absolute zero temperature, there is no random thermal motion of the molecules and all crystalline substances have perfect order. Thus for entropy of the ith specie at its partial pressure p_i and temperature T, we write

$$\tilde{s}(p_i,T) = \tilde{s}^0(T) - R_0 \ln p_i \qquad (10.3)$$

where $\tilde{s}_i^0(T)$ is the absolute specific molar entropy of the ith specie at temperature T and 1 atm. pressure. The value of the partial pressure p_i in Eq. 10.3 is measured in atmospheres. The partial pressure p_i for an ideal gas was defined in Chapter 1 and is given as $p_i = x_i p$, where x_i is the mole fraction of the ith species and p is the total pressure of the mixture in atmospheres. We can therefore write Eq. 10.3 as

$$\tilde{s}_i(T,p_i) = \tilde{s}_i^0(T) - R_0 \ln(x_i p) \qquad (10.4)$$

The absolute entropy of a variety of ideal gases over a range of temperatures and at 1 atm. pressure is available in literature. Equations 10.3 and 10.4 are used to obtain the entropy of species in chemically reacting systems.

10.4   ENTHALPY CHANGES

The enthalpy changes in chemically reacting systems at the reference state can be obtained from changes of the standard heats of formation between the products and the reactants once the chemical reaction is specified and the concentration of the products and the reactants are known. In general, the product species concentrations at given values of pressure and temperature are not known since the combustion may not be complete at the specified pressure and temperature. Thus, there remains the task of finding the equilibrium concentrations of the chemical species of a mixture at given pressure p and temperature T.

10.5   PRODUCT SPECIES IN A CHEMICAL REACTION AT A GIVEN TEMPERATURE AND PRESSURE

The criterion for chemical equilibrium is defined by the minimum value of the Gibbs-free energy. The state of a mixture is specified by (T, p, n_i) where n_i's are the moles of the species of the mixture under equilibrium conditions at temperature T and pressure p. The Gibbs-free energy is given by $G(T,p,n_i)$ and the differential change in the Gibbs-free energy for this state (T,p,n_i) is given by

$$dG = \left(\frac{\partial G}{\partial T}\right)_{p,n_i} dT + \left(\frac{\partial G}{\partial p}\right)_{T,n_i} dp + \sum_i \left(\frac{\partial G}{\partial n_i}\right)_{T,p,n_j,j\neq i} dn_i \qquad (10.5)$$

But $dG = 0$ at equilibrium when G takes on its minimum. At constant (p, T), we can write Eq. 10.5 as

$$dG = \sum \left(\frac{\partial G}{\partial n_i} \right)_{T,p,n_j,j \neq i} dn_i = 0 \tag{10.6}$$

But $\dfrac{\partial G}{\partial n_i} = \tilde{g}_i$, where $\tilde{g}_i$ is the specific molar Gibbs-free energy of the ith specie (per mole of it). This was defined in the previous chapter. Thus at constant p and T, we have for the n_i moles of specie i when the total number of species is N that is, i varying from 1 to N:

$$G = \sum_{i=1}^{N} \tilde{g}_i n_i \tag{10.7}$$

At equilibrium, G is a minimum. The values of n_i are not independent, but are governed by the stoichiometric coefficients of the species and the number of species in the reaction. The number of atoms of the elements in the chemical reaction is, however, conserved and this becomes a constraint. With N species in the reaction, the value of $i = 1, 2, ..., N$. As an example, in the reaction $H_2 + {}^1/_2 O_2 = H_2O$, $N = 3$ with $n_{H_2} = 1$, $n_{O_2} = {}^1/_2$ and $n_{H_2O} = 1$. The number of elements is two, these being H and O.

If in a reaction having N species there are M atoms, we have denoting each of the atom as j, $j = 1, 2, ..., M$, the equations for the conservation of each of the M atoms in the reaction can be written as

$$\sum_{i=1}^{N} v_{j,i} \, n_i - b_j^0 = 0; j = 1,2,...,M \tag{10.8}$$

Here, $v_{j,i}$ represents the atoms of element j in one mole of specie i and b_j^0 is the total number of atoms of the jth element in the reaction.

Denoting $\displaystyle\sum_{i=1}^{N} v_{j,i} \, n_i = b_j$ for $j = 1,2,...M$,we get the equation for the constraint for conserving the atoms as

$$b_j - b_j^0 = 0 ; \quad j = 1,2,...,M$$

In order to minimize $G = \displaystyle\sum_{i=1}^{N} \tilde{g}_i n_i$ subject to the constraint $b_j - b_j^0 = 0$, we multiply the equation for the constraint by Lagrange multipliers λ_j and define a quantity L as

$$L = G + \sum_{j=1}^{M} \lambda_j \left(b_j - b_j^0 \right)$$

Substituting the value of G from Eq. 10.7 and b_j from Eq. 8 in the above expression and differentiating, we get

$$\delta L = \sum_{i=1}^{N}\left(\tilde{g}_i + \sum_{j=1}^{M}\lambda_j v_{j,i}\right)dn_i + \sum_{j=1}^{M}\left(b_j - b^0{}_j\right)d\lambda_j = 0; \text{for } i = 1, 2,\ldots,N$$

Since dn_i and $d\lambda_j$ are not zero being arbitrary independent variables, we have

$$\tilde{g}_i + \sum_{j=1}^{M}\lambda_j v_{j,i} = 0 \quad \text{for } i = 1, 2,\ldots,N$$

(10.9)

$$\text{and } \sum_{i=1}^{N} v_{j,i}n_i - b^0{}_j = 0 \quad \text{for } j = 1,2,\ldots,M$$

The above set of equations give the values of $v_{j,i}$. Thus the equilibrium composition can be determined from the known values of $\tilde{g}_i$ at the specified temperature and pressure. This procedure is used in the different computer codes to determine the equilibrium composition.

Instead of the above method of Lagrange multipliers to minimize the Gibbs-free energy with the constraint that the atoms of the elements must be conserved, the minimization of Gibbs-free energy could also be done using the method of equilibrium constants as given in the following.

From the definition of Gibbs-free energy,

$$\tilde{g}_i = \tilde{h}_i - T\tilde{s}_i = \tilde{h}_i - T\left(\tilde{s}_i^{\circ}(T) - R_0 \ln p_i\right)$$

$$= \tilde{h}_i - T\tilde{s}_i^{0}(T) + R_0T \ln p_i = \tilde{g}_i^{0}(T) + R_0T \ln p_i$$

where $\tilde{g}_i^{0}(T)$ is the temperature-dependent part of $\tilde{g}_i(p,T)$ while the pressure-dependent part is given by $R_0 \ln p_i$.

As an example consider the reaction

$$H_2 + {}^1/_2 O_2 = H_2O$$

(10.10)

We find that when 1 mole of hydrogen disappears, ½ mole of oxygen also vanishes and 1 mole of water is formed. Thus, the change in moles dn_i is related by $dn_{H_2} = -1$, $dn_{O_2} = -1/2$ and $dn_{H_2O} = 1$.

In general for any chemical reaction between reactants A and B to form products C and D, we could write

$$\gamma_A A + \gamma_B B = \gamma_C C + \gamma_D D$$

(10.11)

where $\gamma_A, \gamma_B, \gamma_C, \gamma_D$ are the stoichiometric coefficients. The change in the moles dn_i's is related as

$$-\frac{dn_A}{\gamma_A} = -\frac{dn_B}{\gamma_B} = +\frac{dn_C}{\gamma_C} = +\frac{dn_D}{\gamma_D} = \chi \qquad (10.12)$$

where χ is defined as the degree of the reaction, $0 \leq \chi \leq 1$ ($\chi = 0$ is unreacted and $\chi = 1$ signifies reaction is completed). Thus,

$$dn_A = -\gamma_A \chi, \quad dn_B = -\gamma_B \chi, \quad dn_C = +\gamma_C \chi, \quad dn_D = +\gamma_D \chi \qquad (10.13)$$

and at equilibrium, we write

$$\sum \tilde{g}_i dn_i = \sum_i \tilde{g}_i (\gamma_i \chi) = 0 \qquad (10.14)$$

γ_i is negative for reactants and is positive for products. Since $\gamma_i \chi$ is arbitrary, we have at equilibrium from Eq. 10.14,

$$\sum_i (\tilde{g}_i{}^0 + R_0 T \ln p_i) \gamma_i \chi = 0 \qquad (10.15)$$

The above equation involving the minimization of Gibbs-free energy is used to determine the species formed in equilibrium in a chemical reaction at the given pressure and temperature.

We can express Eq. 10.15 for the general reaction given by Eq. 10.11 as

$$\left(-\gamma_A \tilde{g}_A{}^0 - \gamma_B \tilde{g}_B{}^0 + \gamma_C \tilde{g}_C{}^0 + \gamma_D \tilde{g}_D{}^0\right) + R_0 T \ln \frac{p_C{}^{\gamma_C} p_D{}^{\gamma_D}}{p_A{}^{\gamma_A} p_B{}^{\gamma_B}} = 0 \qquad (10.16)$$

Defining

$$\Delta G^0(T) = -\gamma_A \tilde{g}_A{}^0 - \gamma_B \tilde{g}_B{}^0 + \gamma_C \tilde{g}_C{}^0 + \gamma_D \tilde{g}_D{}^0 \qquad (10.17)$$

Equation 10.16 can be written as

$$\ln \frac{p_C{}^{\gamma_C} p_D{}^{\gamma_D}}{p_A{}^{\gamma_A} p_B{}^{\gamma_B}} = -\frac{\Delta G^0(T)}{R_0 T} = f(T) \qquad (10.18)$$

Denoting the pressure ratios on the left side as an equilibrium constant K at temperature T, we have

$$K(T) = \frac{p_C{}^{\gamma_C} p_D{}^{\gamma_D}}{p_A{}^{\gamma_A} p_B{}^{\gamma_B}} = \frac{x_C{}^{\gamma_C} x_D{}^{\gamma_D}}{x_A{}^{\gamma_A} x_B{}^{\gamma_B}} p^{\gamma_C + \gamma_D - \gamma_A - \gamma_B} \qquad (10.19)$$

where for ith specie, the molar concentration is $x_i = \frac{n_i}{n}$ and the partial pressure is $p_i = x_i p$ from Dalton's law of partial pressures. Here, the total number of moles of the mixture is n and the total pressure is p. The equilibrium constant $K(T)$ is a function of temperature and is determined from the change in the Gibbs-free energy at the standard pressure and temperature T, namely, that given in Eq. 10.17 divided by R_0T for the reaction. It may be kept in mind that the calculation of species is based on the assumption that the reactants and products are ideal gases.

The equilibrium composition varies with temperature. The Gibbs-free energy $G = H - TS$ gives the incremental change $dG = dH - TdS - SdT$. For an ideal gas at constant pressure $dH = TdS$ giving $dG = -SdT$ and

$$\frac{dG}{dT} = -S$$

Substituting the value of S in the expression for G, we get

$$G = H + T\frac{dG}{dT}$$

Dividing the above by T^2 and simplifying, we can express the equation as

$$\frac{d}{dT}\left(\frac{G}{T}\right)_p = -\frac{H}{T^2}$$

and can be further written as

$$\frac{d}{dT}\left(\frac{\Delta G}{T}\right)_p = -\frac{\Delta H}{T^2} \tag{10.20}$$

Combining with Eq. 10.18 at the standard pressure of 1 atmosphere, we get

$$\frac{d}{dT}\left(\ln\ K(T)\right) = \frac{\Delta H}{R_0T^2} \tag{10.21}$$

The above is known as the van't Hoff equation and is used in calculating the equilibrium constants with changes in temperature at the standard pressure of 1 atm.

10.6   EXAMPLE OF DETERMINING EQUILIBRIUM COMPOSITION

Consider a mixture CO_2, O_2 and N_2 in volumetric proportion 1:½:½ at 3,000 K and 1 atm. pressure for which we require to determine the equilibrium composition. We write the overall composition to be given by the products of the reaction

$$CO_2 + \tfrac{1}{2}\ O_2 + \tfrac{1}{2}\ N_2 = aCO + bNO + cCO_2 + dO_2 + eN_2 \tag{10.22}$$

where the moles a, b, c, d and e for CO, NO, CO_2, O_2 and N_2 are not known. From the conservation of the atoms in the above reaction, we write

C atoms: $a + c = 1$
O atoms: $a + b + 2c + 2d = 3$
N atoms: $b + 2e = 1$

Solving for c, d and e in terms of a and b, we get $c = 1 - a$, $d = 1 + (a-b)/2$ and $e = (1-b)/2$. We still need two more equations to solve for a, b, c, d and e and thus determine the equilibrium concentration of the species.

The total number of moles of the product is given by

$$a + b + c + d + e = (4 + a)/2$$

Thus, the mole fractions for the various species are obtained as

$$x_{CO} = \frac{2a}{4+a}$$

$$x_{NO} = \frac{2b}{4+a}$$

$$x_{CO_2} = \frac{2(1-a)}{4+a}$$

$$x_{O_2} = \frac{1+a-b}{4+a}$$

$$x_{N_2} = \frac{1-b}{4+a}$$

Two reactions among the species that provide the additional two equations are

$$\text{Reaction 1}: CO_2 \Leftrightarrow CO + \frac{1}{2}O_2 \tag{10.23}$$

$$\text{Reaction 2}: \frac{1}{2}O_2 + \frac{1}{2}N_2 \Leftrightarrow NO \tag{10.24}$$

The equilibrium constant K is determined from Eqs. 10.18 and 10.19 as

$$\ln(K(T)) = -\frac{\Delta G^0(T)}{R_0 T}.$$

In the case of reaction 1 (Eq. 10.23), the value of $\Delta G^0(T)$ is given at $T = 3000$ K as $\Delta G^0(3000) = \tilde{g}_{CO}(3000) + \frac{1}{2}\tilde{g}_{O2}(3000) - \tilde{g}_{CO2}(3000)$. Using the values of the molar Gibbs function $\tilde{g}$ for the different species CO, O_2 and CO_2 available in literature, at different temperatures and 1 atm. pressure, we get at $T = 3000$ K

$$K_1 = 0.3273$$

Similarly the equilibrium constant for the Reaction 2 at 3000 K is obtained as

$$K_2 = 0.1222$$

Substituting the values in the equilibrium constants for Reactions 1 and 2 (Eqs. 10.23 and 10.24), respectively, we get

$$K_1 = 0.3273 = \frac{x_{CO}x_{O_2}^{1/2}}{x_{CO_2}} p^{1+\frac{1}{2}+1} \text{ giving}$$

$$\left(\frac{a}{1-a}\right)\left(\frac{1+a-b}{4+a}\right)^{1/2} = 0.3273 \tag{10.25}$$

$$K_2 = 0.1222 = \frac{x_{NO}}{x_{O_2}^{1/2}x_{N_2}^{1/2}} p^{1-\frac{1}{2}-\frac{1}{2}} \text{ or} \left(\frac{2b}{4+a}\right)\left(\frac{(4+a)^2}{(1+a-b)(1-b)}\right)^{1/2} \text{ giving}$$

$$\frac{2b}{\left[(1-b)(1+a-b)\right]^{1/2}} = 0.1222 \tag{10.26}$$

Solving for a and b from Eqs. 10.25 and 10.26 gives

$$a = 0.3745, b = 0.0675$$

Expressing c, d, and e in terms of a and b gives $c=0.6255$, $d=0.6535$ and $e=0.4663$.

10.7 CHEMICAL EQUILIBRIUM OF SPECIES AT GIVEN TEMPERATURE AND VOLUME

For chemical reactions taking place at constant volume, the Helmholtz function (A) is used instead of the Gibbs-free energy. From the previous chapter on thermodynamic equilibrium, we have

$$dA(T,V) = -SdT - pdV + \left(\frac{\partial A}{\partial n_i}\right)_{V,T,n_{j,\neq i}} dn_i$$

At equilibrium $dA=0$ and A is a minimum. We derive an equation similar to Eq. 10.15 for a given temperature at constant volume as

$$\sum_i (\tilde{a}_i^0(T) + R_0 T \ln \tilde{v}_i)\gamma_i \chi = 0 \tag{10.27}$$

where $\tilde{a}_i$ is the specific Helmholtz function per mole for specie i. The partial volume of specie i is $\tilde{v}_i$ per mole and is defined by Amagat's law for partial volumes for an ideal gas mixtures (Chapter 2). $\tilde{a}_i^0(T)$ denotes the temperature dependent part of Helmholtz function. The procedure in the last section is followed by replacing Gibbs-free energy change $\Delta G^0(T)$ in Eq. 10.18 by the change of the Helmholtz function $\Delta A^0(T)$ for the reaction and defining the equilibrium constant $K(T)$ at constant volume. The species are determined at a given specified temperature and volume.

10.8 CORRECTIONS FOR REAL GAS: FUGACITY

The determination of the concentration of species under equilibrium assumed them as ideal gases. At higher values of pressures and lower temperatures, the ideal gas assumption may not be valid. The deviation from the idealized gas assumption is expressed by the pressure dependence of the Gibbs-free energy of a real gas by an effective pressure known as fugacity.

For a single component system of real gas

$$d\mu = d\tilde{g} = -\tilde{s}dT + \tilde{v}dp \qquad (10.28)$$

where $\tilde{g}$, $\tilde{s}$ and $\tilde{v}$ denote, as before, the specific Gibbs-free energy, specific entropy and specific volume all per mole. At constant temperature, the change of chemical potential from a sufficiently low pressure p_0 to a higher value of pressure p is therefore

$$\int_{\mu_0}^{\mu} d\mu = \int_{p_0}^{p} \tilde{v}\,dp \qquad (10.29)$$

where μ and μ_0 correspond to the chemical potential at p and p_0, respectively.

For an ideal gas, the specific volume per mole is $\tilde{v} = \dfrac{R_0 T}{p}$ and $\mu - \mu_0 = R_0 T \ln \dfrac{p}{p_0}$. However, for a dense gas, the specific volume and chemical potential such as at the higher values of pressure p cannot be approximated by the expressions corresponding to ideal gas. The fugacity f is defined as the corrected value of pressure so that the chemical potential is the same as that in the real gas

$$\mu - \mu_0 = R_0 T \ln \frac{f}{f_0} \qquad (10.30)$$

where f=fugacity when pressure is p and f_0 when the pressure is p_0. The use of fugacity provides a fundamental way of determining equilibrium of chemical reactions in real gases. Equation 10.30 can be integrated to give

$$f = f_0 \exp\left(\frac{\mu - \mu_0}{R_0 T} \right)$$

The Gibbs-free energy of a real gas following Eq. 10.30 can be written as

$$\tilde{g} = \tilde{g}_0 + R_0 T \ln \frac{f}{f_0} \qquad (10.31)$$

The relation between fugacity f and the true value of pressure p is expressed as

$$f = \phi p \qquad (10.32)$$

where ϕ is the dimensionless fugacity coefficient which depends on temperature, pressure and the gas species.

If the value of Gibbs-free energy per mole $\tilde{g}$ is substituted for the chemical potential μ in the integrated form of Eq. 10.28, we get

$$\tilde{g}(p) = \tilde{g}(p_0) + \int_p^{p_0} \tilde{v}dp \tag{10.33}$$

which is valid for all gases whether ideal or real.

Substituting the form of Eq. 10.31 for fugacity in Eq. 10.33

$$\int_{p_0}^p \tilde{v}dp = \tilde{g}(p) - \tilde{g}(p_0) = R_0T \ln \frac{f}{f_0} \tag{10.34}$$

If the gas is ideal and its specific volume per mole is denoted by $\tilde{v}_i$

$$\int_{p_0}^p \tilde{v}_i dp = R_0T \ln \frac{p}{p_0} \tag{10.35}$$

The difference between Eqs. 10.34 and 10.35 is

$$\int_{p_0}^p (\tilde{v} - \tilde{v}_i)dp = R_0T \ln\left[\frac{f/f_0}{p/p_0}\right]$$

which can be rearranged to give

$$\ln\left[\frac{f/p}{f_0/p_0}\right] = \frac{1}{R_0T} \int_{p_0}^p (\tilde{v} - \tilde{v}_i)dp \tag{10.36}$$

When $p_0 \to 0$ where the gas is ideal, we have

$$\frac{f_0}{p_0} \to 1 \text{ as } p_0 \to 0$$

Under the above condition for the low pressures p_0,

$$\ln(f/p) = \frac{1}{R_0T} \int_0^p (\tilde{v} - \tilde{v}_i)dp \tag{10.37}$$

With $\phi = \dfrac{f}{p}$

$$\ln\phi = \frac{1}{R_0T} \int_0^p (\tilde{v} - \tilde{v}_i)dp \tag{10.38}$$

Since for an ideal gas

$$\tilde{v}_i = \frac{R_0T}{p}$$

while for a real gas

$$\tilde{v} = \frac{Z R_0 T}{p}$$

where Z is the compressibility factor, we get

$$\ln \phi = \int_0^p \frac{Z-1}{p} dp \qquad (10.39)$$

We have seen the variation of Z with pressure in the generalized compressibility chart in the Chapter 2 on equation of state. The fugacity coefficient can thus be determined and the fugacity related to the actual pressure. In general, over the range of pressures for which $Z < 1$, that is, up to moderate values of pressure

$$f < p$$

and the Gibbs energy per mole is less than that of an ideal gas at the given value of pressure.

However, for $Z > 1$, which takes place at sufficiently high pressures, $\phi > 1$ and

$$f > p$$

with the Gibbs-free energy being greater than that of an ideal gas.

The temperature dependence of fugacity can be obtained from Eq. 10.31 by substituting $p_0 = f_0$ since at the low pressure limit the gas behaves as an ideal gas and

$$\Delta \tilde{g} = R_0 T \ln \frac{f}{p_0}$$

where the fugacity f corresponds to the high pressure gas. Using Eq. 10.20, we get

$$\frac{d}{dT} \left(\frac{\Delta \tilde{g}}{T} \right)_p = -\frac{\Delta \tilde{h}}{T^2}$$

Hence $\qquad \left(\frac{d \ln f}{dT} \right)_p = -\frac{\Delta \tilde{h}}{R_0 T^2} = \frac{\tilde{h}_i - \tilde{h}}{R_0 T^2} \qquad (10.40)$

where $\tilde{h}_i$ is the enthalpy per mole of the gas in the ideal gas limit and is independent of the temperature while $\tilde{h}$ is the enthalpy of the real gas.

However,

$$\tilde{h}_i - \tilde{h} = \int_0^p \left(\frac{\partial \tilde{h}}{\partial p} \right)_T dp = \int_0^p \left(\mu_{JT} \tilde{c}_p dp \right)$$

where μ_{JT} is the Joule Thomson coefficient. The symbol μ_π is used to distinguish it from the chemical potential μ.

Thus,

$$\left(\frac{\partial \ln f}{\partial T}\right)_p = \frac{\displaystyle\int_0^p \mu_{JT}\tilde{c}_p \, dp}{R_0 T^2} \tag{10.41}$$

The fugacity f so determined is used in place of pressure for estimating the equilibrium of the species in the chemical reaction.

11 Statistical Thermodynamics

11.1 INTRODUCTION

Thermodynamics does not acknowledge the molecular structure of matter. However, to acquire a better understanding of thermodynamics, it is necessary to discuss the molecular structure of matter and the molecular basis of thermodynamics. Microscopic description of the behavior of atoms and molecules, referred to as particles, is based on quantum mechanics.

The number of particles in a macroscopic system is very large, $O[10^{23}]$ particles. Thus, macroscopic thermodynamic variables are averaged over the microscopic variables describing the molecules. This is the subject of statistical thermodynamics and provides the formalism to link the microscopic variables to the macroscopic thermodynamic variables like internal energy, entropy, temperature, pressure, etc.

It can be said that the central problem in statistical thermodynamics is to determine the equilibrium distribution of the particles among their accessible quantum states. It is assumed that the equilibrium distribution corresponds to the most probable distribution. The most probable distribution is considered to be the distribution that corresponds to the maximum number of ways in which the distribution can be realized subject to the macroscopic constraints of the system. Thus, the first step is to determine the number of ways to realize a given distribution of the particles among their accessible quantum states (or equivalently their energy levels).

11.2 DISTRIBUTION OF PARTICLES AND THEIR ENERGY LEVELS: BOSE–EINSTEIN, FERMI–DIRAC AND BOLTZMANN STATISTICS

Particles can be classified, in general, as Bosons and Fermions. For Bosons, there are no restrictions on the number of particles that can occupy a quantum state. However, for Fermions, the exclusion principle restricts that only one particle can occupy a given state. Since the number of particles in a given energy level is generally very small compared to the degeneracy of the energy levels, it is highly unlikely that more than one particle would occupy a single degenerate state. Thus, the distribution of particles for the Bosons and Fermions are essentially the same.

Let us first consider the Bosons. We would like to determine the number of ways to distribute a total of N particles so that we have N_1 particles in energy level ε_1, N_2 particles in energy level ε_2, ..., N_i particles in energy level ε_i, N_j particles in energy level ε_j etc. The degeneracy of energy level ε_i is g_i, that is, number of quantum states in energy level ε_i. The question is to determine the many ways to distribute N_i particles among g_i quantum states corresponding to energy level ε_i.

DOI: 10.1201/9781003224044-11

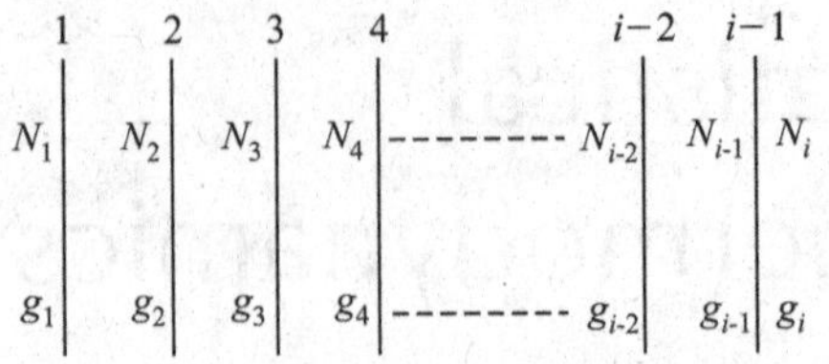

FIGURE 11.1 g_i Boxes with $(i-1)$ partitions.

Consider the g_i states to be g_i number of boxes to distribute N_i particles. Let the g_i boxes be defined by $i-1$ partitions as illustrated in Figure 11.1.

Thus, a distribution consists of specifying the number of particles in each box, that is, N_1 particles in box g_1, N_2 particles in box g_2, ..., N_i particles in box g_i, etc.

Assuming that the particles and the partitions to be distinguishable for the moment, the number of ways to arrange $(g_i - 1 + N_i)$ distinguishable objects would be $(g_i - 1 + N_i)!$. However, the particles as well as the partitions are indistinguishable, that is, interchanging the partitions and the particles leave the distribution invariant. We therefore have counted $(g_i - 1)!$ partitions and $N_i!$ way too much. Thus, the number of ways to realize the distribution of N_i distinguishable particles among g_i states in energy level ε_i would be

$$W_i = \frac{(g_i - 1 + N_i)!}{N_i!(g_i - 1)!}$$

Similarly the number of ways to distribute N_j indistinguishable particles among g_j quantum states at energy level ε_j gives

$$W_j = \frac{(N_j + g_j - 1)!}{N_j!(g_j - 1)!}$$

Since each particular combination in the ε_i level can be combined with the combinations in the ε_j level, the total number of ways (or permutations) to distribute N_i particles in energy levels ε_i and N_j particles in ε_j would be $W_i W_j$.

Generalizing, we obtain

$$W = \prod_i \frac{(g_{i-1} + N_i)!}{(g_i - 1)! N_i!} \tag{11.1}$$

for the total number of ways to distribute N particles among their accessible energy levels. The symbol Π denotes the product of i similar terms.

If $g_i \gg N_i$, the general permutation formula given in Eq. 11.1 can be reduced to a simple expression as follows. First, we write

$$(g_i + N_i - 1)! = (g_i - 1 + N_i)(g_i - 1 + N_i - 1)(g_i - 1 + N_i - 2)$$

$$----- (g_i - 1 + N_i - (N_i - 1))g_i(g_i - 1)(g_i - 2)$$

$$---- (g_i - (g_i - 1)) \approx g_i^{n_i}(g_i - 1)!$$

where we have approximated the g_i terms as $g_i^{N_i}$. Thus, Eq. 11.1 can be written as

$$W = \prod \frac{(N_i + g_i - 1)!}{N_i!(g_i - 1)!} = \prod \frac{g_i^{N_i}(g_i - 1)!}{N_i!(g_i - 1)!}$$

$$\text{or } W = \prod \frac{g_i^{N_i}}{N_i!} \tag{11.2}$$

Thus, the total number of ways to realize a distribution of N_i particles in energy level ε_i, N_j in energy level ε_j, ... etc., is given by Eq. 11.2.

In the derivation of Eqs. 11.1 and 11.2, no restrictions are stated for the number of partitions or for particles that can occupy a quantum state. Thus, Eq. 11.1 is valid for Bosons. For Fermions, only one particle is permitted per quantum state. Hence, it is necessary that $g_i > N_i$. To obtain the number of ways to realize a given distribution of Fermions, we reason as follows:

Consider the ε_i energy level with g_i states; it is desirous that $g_i > N_i$. For the first particle, there are g_i states to choose from. Once the choice is made for the first particle, there are $(g_i - 1)$ choices left for the second particle, and similarly $(g_i - 2)$ choices for the third particle. For the N_i^{th} particle, there are $(g_i - N_i + 1)$ choices. Since each of the g_i choice for the first particle can go with $(g_i - 1)$ choices for the second particle and $(g_i - 2)$ choices for the third and so on, we see that for N_i particles there will be a total of $g_i (g_i - 1) (g_i - 2)----(g_i - N_i + 1)$ choices. Since the N_i particles are indistinguishable, interchanging particles among the g_i compartments leave the distribution invariant and we have counted $N_i!$ too much. Thus, the distribution of N_i particles in the g_i degenerate states of the energy level ε_i is

$$W_i = \frac{g_i(g_i - 1)(g_i - 2)----(g_i - N_i + 1)}{N_i!}$$

The above can be reduced to a more convenient form by writing the numerator in the above expression as

$$\frac{g_i(g_i - 1)(g_i - 2)----(g_i - N_i + 1)(g_i - N_i)(g_i - N_i - 1)----((g_i - N_i)-(g_i - N_i - 1))}{(g_i - N_i)(g_i - N_i - 1)---((g_i - N_i)-(g_i - N_i - 1))}$$

$$= \frac{g_i!}{(g_i - N_i)!}$$

We therefore obtain

$$W_i = \frac{g_i!}{(g_i - N_i)!N_i!} \tag{11.3}$$

and for a total of N Fermions among their accessible states for the various energy levels, we obtain the distribution for Fermions to be given by

$$W = \prod_i \frac{g_i!}{(g_i - N_i)! N_i!} \qquad (11.4)$$

Equation 11.4 is to be compared with Eq. 11.1 for Bosons. In the case of Bosons, if $g_i \gg N_i$, we write

$$g_i! = g_i(g_i - 1)(g_i - 2) - - - - - (g_i - N_i + 1)(g_i - N_i)!$$

giving

$$g_i! = g_i^N (g_i - N_i)!$$

Thus, Eq. 11.4 reduces to

$$W = \prod_i \frac{g_i^{N_i}}{N_i!}$$

which is the same result as Eq. 11.2. The fact that the permutation formula for both Fermions and Bosons reduces to the same limit for $g_i \gg N_i$, that is, there are numerous states for the N_i particles to occupy. Thus, the probability of particles crowding into one state is negligible and most of the g_i states are unoccupied. Hence whether the restriction as to the number of particles permitted per quantum state is imposed or not makes little difference when $g_i \gg N_i$.

The permutation equation for Boson given by Eq. 11.1 is referred to as Bose-Einstein statistics, whereas Eq. 11.4 is known as Fermi-Dirac statistics. In the limit $g_i \gg N_i$, Eq. 11.2 is referred to as Boltzmann statistics.

11.3 MAXWELL–BOLTZMANN DISTRIBUTION: PARTITION FUNCTION

It is reasonable to assume that at equilibrium, the distribution corresponds to the most probable one. We shall consider Boltzmann statistics given by Eq. 11.2 and maximize W. In other words, we seek the distribution that maximizes the number of ways to realize the distribution. Since $\ln W$ is a monotonic function of W, it is more convenient to maximize $\ln W$, that is,

$$\ln W = \sum N_i \ln g_i - \sum N_i \ln N_i + \sum N_i \qquad .$$

where we have used Stirling's formula for $\ln N! = N \ln N - N$.

Taking the variation of $\ln W$, we write

$$\delta \ln W = \sum_i \ln g_i \delta N_i - \sum_i \ln N_i \delta N_i - \sum_i \delta N_i + \sum_I \delta N_i$$

$$= \sum_i \ln \frac{g_i}{N_i} \delta N_i$$

The δN_i's are not independent in the above equation, but subject to the constraint $\sum N_i = N = \text{constant}$ and $\sum N_i \varepsilon_i = U = \text{constant}$. Taking the variation of the constraints, we obtain $\sum \delta N_i = 0$ and $\sum \varepsilon_i \delta N_i = 0$. Multiplying by the yet undermined multipliers (Lagrange multipliers) α and β, that is, $\sum \alpha \delta N_i = 0$ and $\sum \beta \varepsilon_i \delta N_i = 0$ and subtracting them from $\delta \ln W$, we get

$$\delta \ln W = \sum \left(\ln \frac{g_i}{N_i} - \alpha - \beta \varepsilon_i \right) \delta N_i = 0$$

The δN_i's are now independent and thus we write

$$\ln \frac{g_i}{N_i} - \alpha - \beta \varepsilon_i = 0$$

$$N_i^0 = g_i e^{-\alpha} e^{-\beta \varepsilon_i} \tag{11.5}$$

where we write the superscript N_i^0 to denote the equilibrium distribution.

Using the constraint $\sum N_i = N$, we obtain

$$\sum N_i^0 = N = e^{-\alpha} \sum g_i e^{-\beta \varepsilon_i}$$

Thus, Eq. 11.5 becomes

$$N_i^0 = N \frac{g_i e^{-\beta \varepsilon_i}}{z} \tag{11.6}$$

where

$$z = \sum_i g_i e^{-\beta \varepsilon_i} \tag{11.7}$$

is called the partition function. It expresses the distribution or partition of energies over various energy levels.

Equation 11.6 is referred to as the Maxwell–Boltzmann distribution and is perhaps the most important formula in statistical thermodynamics.

To determine the other multiplier β, consider an isolated system consisting of two systems A and B separated by a diathermal wall to permit exchange of heat. This is shown in Figure 11.2. The combined system $A + B$ is an isolated system.

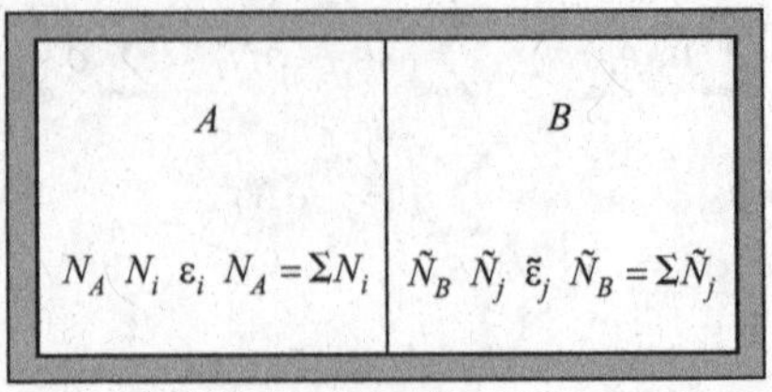

FIGURE 11.2 Systems A and B exchanging heat.

For systems A and B, we write

$$W_A = \prod_i \frac{g_i^{N_i}}{N_i!}$$

and similarly for B, we write

$$W_B = \prod_i \frac{\tilde{g}_i^{\tilde{N}_i}}{\tilde{N}_i!}$$

For the combined system $A + B$, the number of ways to realize a given distribution of N_i particles among energy levels ε_i in A and $\tilde{N}_j$ particles among levels $\tilde{\varepsilon}_j$ in B is given by $W_{AB} = W_A \times W_B$ giving

$$\ln W_{AB} = \ln W_A + \ln W_B$$

$$= \sum N_i \ln g_i - \sum N_i \ln N_i + \sum N_i + \sum \tilde{N}_j \ln \tilde{g}_j - \sum \tilde{N}_j \ln \tilde{N}_j + \sum \tilde{N}_j$$

Taking the variation of W_{AB}, we obtain

$$\delta \ln W_{AB} = \sum \ln \frac{g_i}{N_i} \delta N_i + \sum \ln \frac{\tilde{g}_j}{\tilde{N}_j} \delta \tilde{N}_j$$

The constraints for the above are

$$\sum N_i = N_A = \text{constant}$$

$$\sum \tilde{N}_i = \tilde{N}_B = \text{constant}$$

$$\sum N_i \varepsilon_i + \sum \tilde{N}_j \tilde{\varepsilon}_j = U = \text{constant}$$

Here, U is the total energy of the combined system $A + B$ and is a constant. Note that because of heat exchange between A and B, we cannot specify the internal energy of A or B individually. However, the internal energy of the combined system $A + B$ is $U = $ constant for the isolated system. Taking the variation of the constraints and multiplying by $-\alpha, -\tilde{\alpha}$ and $-\beta$, that is,

$$\sum -\alpha \delta N_i = 0$$

$$\sum -\tilde{\alpha} \delta \tilde{N}_j = 0$$

$$\sum -\beta \varepsilon_i \delta N_i - \beta \tilde{\varepsilon}_j \delta \tilde{N}_j = 0$$

and adding to the expression for $\delta \ln W_{AB}$, we obtain

$$\delta \ln W_{AB} = \sum_i \left(\ln \frac{g_i}{N_i} - \alpha - \beta \varepsilon_i \right) \delta N_i + \sum_j \left(\ln \frac{\tilde{g}_j}{\tilde{N}_j} - \tilde{\alpha} - \beta \tilde{\varepsilon}_j \right) \delta \tilde{N}_j = 0$$

The δN_i's and $\delta \tilde{N}_j$'s are now made independent via the Lagrangian multipliers. Thus, we write the equilibrium distribution for systems A and B as

$$N_i^0 = g_i e^{-\alpha} e^{-\beta \varepsilon_i}$$

$$\tilde{N}_i^0 = \tilde{g}_i e^{-\tilde{\alpha}} e^{-\beta \tilde{\varepsilon}_i}$$

The Lagrangian multipliers α and $\tilde{\alpha}$ which arise for the constraint of N_i and $\tilde{N}_i$ in subsystems A and B are to be related to N_A and $\tilde{N}_B$. The multiplier β is based on the constraint of energy and is related to the general levels of energy.

Normalizing the above relations, we have

$$N_i^0 = N_A \frac{g_i e^{-\beta \varepsilon_i}}{z} \tag{11.8}$$

$$\tilde{N}_j^0 = \tilde{N}_B \frac{\tilde{g}_i e^{-\beta \tilde{\varepsilon}_j}}{\tilde{z}} \tag{11.9}$$

where the partition functions

$$z = \sum_i g_i e^{-\beta \varepsilon_i}$$

$$\tilde{z} = \sum_j \tilde{g}_j e^{-\beta \tilde{\varepsilon}_j} \tag{11.10}$$

The sum is taken over i and j quantum states. The partition functions z and $\tilde{z}$ are related to the equilibrium distribution.

The above result indicates that for the two systems separated by a diathermal wall, their equilibrium distributions share the same value of the parameter β. Thus, the parameter must be equivalent to the temperature of the systems. We shall define

$$\beta = \frac{1}{kT}$$

where k is a constant with dimensions energy/degree (e.g., $k = 1.381 \times 10^{-23}$ J/K).

11.4 BOLTZMANN'S FORMULA

We note that the fundamental postulate of statistical thermodynamics is that at equilibrium, $\ln W$ is a maximum. Further for two systems in thermal contact at equilibrium, $W_{AB} = W_A \times W_B$ and hence $\ln W_{AB} = \ln W_A + \ln W_B$, that is, the law has the additive property. This led Boltzmann to formally equate $\ln W$ to the entropy function, that is,

$$S = k \ln W^0 \tag{11.11}$$

The above definition gives maximum entropy at equilibrium and the additive property

$$S_{AB} = S_A + S_B$$

It is to be noted that W^0 corresponds to the equilibrium distribution. Further with

$$d \ln W = \sum_i \ln \frac{g_i}{N_i} dN_i$$

and

$$\ln \frac{g_i}{N_i} = \alpha + \beta \varepsilon_i$$

$$d \ln W = \alpha \sum dN_i + \beta d \sum \varepsilon_i N_i$$

$$= \beta dU$$

as $U = \sum \beta \varepsilon_i$ and $\sum N_i = N$

From thermodynamics,

$$dS = \frac{1}{T} dU + \frac{p}{T} dV$$

$$= \left(\frac{\partial S}{\partial U} \right)_V dU + \left(\frac{\partial S}{\partial V} \right)_U dV$$

and equating the coefficients gives

$$\frac{1}{T} = \left(\frac{\partial S}{\partial U} \right)_V ; \quad \frac{p}{T} = \left(\frac{\partial S}{\partial V} \right)_U \tag{11.12}$$

From Eq. 11.11,

$$\left(\frac{\partial S}{\partial U}\right)_V = k\,\frac{\partial \ln W}{\partial U} = \beta k = \frac{1}{T}$$

Thus,

$$\beta = \frac{1}{kT} \tag{11.13}$$

as indicated previously.

Using the equilibrium distribution given by Eq. 11.6, an expression for entropy as defined by Boltzmann (Eq. 11.11) can be obtained as

$$S = k \ln W = kN \ln \frac{z}{N} + k\beta U + kN$$

With

$$\beta = \frac{1}{kT}$$

the above can be written as

$$S = kN \ln \frac{z}{N} + \frac{U}{T} + kN \tag{11.14}$$

Thus from Eq. 11.12, we can obtain pressure as

$$p = NkT\left(\frac{\partial \ln z}{\partial V}\right)_U \tag{11.15}$$

Since, the internal energy is

$$U = \sum n_i \varepsilon_i = \sum \frac{g_i e^{-\beta \varepsilon_i}}{z} \varepsilon_i$$

and with $z = \sum g_i e^{-\beta \varepsilon_i}$

we can write

$$U = -\frac{N}{z}\left(\frac{\partial z}{\partial \beta}\right)_V = NkT\left(\frac{\partial \ln z}{\partial \ln T}\right)_V \tag{11.16}$$

since $\beta = \dfrac{1}{kT}$

Equations 11.14–11.16 link the thermodynamic functions S, U and p in terms of the partition function z, which is based on the molecular energy levels of the particles in equilibrium.

11.5 PARTITION FUNCTION FOR A MONOATOMIC GAS: INTERNAL ENERGY, PRESSURE, EQUATION OF STATE AND ENTROPY OF AN IDEAL GAS

A monoatomic gas stores energy in the translational modes at the temperatures of interest. At low to moderate pressure and temperatures of interest, it can be considered as an ideal gas. The translational energy level of the individual particles in the monoatomic gas is given by

$$\varepsilon_i = \frac{h^2}{8mV^{2/3}}\left(n_X^2 + n_Y^2 + n_Z^2\right) \tag{11.17}$$

where h is the Planck constant, m is the mass of the particle, V is the volume of the system, and n_x, n_y, n_z are the quantum numbers. From Eq. 11.17, the partition function can be evaluated as

$$z = \sum_{n_x=1}^{\infty}\sum_{n_y=1}^{\infty}\sum_{n_z=1}^{\infty} e^{-\frac{h^2}{8mkTV^{2/3}}\left(n_x^2+n_y^2+n_z^2\right)}$$

Since the summation is over quantum states, we need not consider the degeneracy of the energy levels. The values of the quantum numbers are very large for any appreciable energy, the change of n_x, n_y, n_z are very small and we can replace the summation by an integral, that is,

$$z = \left[\int_0^{\infty} e^{-\frac{h^2}{8mkTV^{2/3}}\left(n_x^2\right)} dn_x\right]\left[\int_0^{\infty} e^{-\frac{h^2}{8mkTV^{2/3}}\left(n_y^2\right)} dn_y\right]\left[\int_0^{\infty} e^{-\frac{h^2}{8mkTV^{2/3}}\left(n_z^2\right)} dn_z\right]$$

Each integral in the above is of the form

$$\int_0^{\infty} e^{-ax^2} dx = \frac{1}{2}\sqrt{\frac{\pi}{a}}$$

We therefore obtain

$$z = V\left(\frac{2\pi mkT}{h^2}\right)^{3/2} \tag{11.18}$$

Hence,

$$\ln z = \ln V + \frac{3}{2}\ln T + \frac{3}{2}\ln\left(\frac{2\pi mk}{h^2}\right)$$

From the expression above, Eq. 11.16 gives

$$U = NkT\left(\frac{\partial \ln z}{\partial \ln T}\right)_V = \frac{3}{2}NkT \tag{11.19}$$

The pressure from Eq. 11.15 becomes

$$p = NkT \left(\frac{\partial \ln z}{\partial V} \right)_U = \frac{NkT}{V} \qquad (11.20)$$

$$\text{or } p = \frac{nR_0 T}{\tilde{v}}$$

which is the equation of state of an ideal gas.

The entropy from Eq. 11.14 can be readily obtained as

$$S = kN \ln \frac{Z}{N} + \frac{U}{T} + kN$$

$$= Nk \left[\frac{3}{2} \ln T + \ln \frac{V}{N} + \ln \left(\frac{2\pi mk}{h^2} \right)^{3/2} + \frac{5}{2} \right] \qquad (11.21)$$

If we consider 1 mole of a gas, that is, $N = N_0 = 6.023 \times 10^{23}$ molecules per mole and $N_0 k = R_0$, Eq. 11.21 becomes

$$\tilde{S} = \tilde{C}_V \ln T + R_0 \ln \tilde{v} + R_0 \ln \frac{1}{N_0} \left(\frac{2\pi mk}{h^2} \right)^{3/2} + \frac{5}{2} R_0 \qquad (11.22)$$

where $\tilde{v} = \dfrac{V}{n}$, $\tilde{c}_v = \dfrac{(3/2)R_0}{n}$, and $n = \dfrac{N}{N_0}$. Eq. 11.22 was first obtained by Sakur and Tetrode.

11.6   REVERSIBLE HEAT TRANSFER, WORK AND THE FIRST LAW

Since the internal energy is given by

$$U = \sum N_i \varepsilon_i$$

we can write the change in internal energy as

$$dU = \sum \varepsilon_i dN_i + \sum N_i d\varepsilon_i$$

The first term $\sum \varepsilon_i dN_i$ on the right hand side denotes a change of internal energy resulting from a change in the distribution of the particles among its energy levels. This term represents reversible heat transfer at constant volume and therefore no change in the energy levels ε_i occurs. Thus, we write

$$dQ_{\text{rev}} = \sum \varepsilon_i dN_i$$

Writing the second term as

$$\sum N_i \frac{d\varepsilon_i}{dV} dV$$

we see that it denotes a change in the energy level ε_i resulting from a change in the volume of the system.

For the particular case of a monoatomic gas, Eq. 11.17 gives the energy level as a function of the volume of the system. We may write

$$\ln \varepsilon_i = -\frac{2}{3}\ln V + \ln\left(n_x^2 + n_x^2 + n_z^2\right) + \ln \frac{h^2}{8m}$$

Thus

$$\frac{d\varepsilon_i}{\varepsilon_i} = -\frac{2}{3}\frac{dV}{V}$$

and

$$N_i d\varepsilon_i = -\frac{2}{3}\frac{N_i\varepsilon_i}{V}dV - \frac{2}{3}\frac{U}{V}dV$$

From Eqs. 11.20 and 11.19, we have for a monoatomic gas

$$p = \frac{NkT}{V}; \quad U = \frac{3}{2}NkT$$

and therefore get

$$p = \frac{2}{3}\frac{U}{V}$$

Thus

$$N_i d\varepsilon_i = -pdV$$

which corresponds to the work done by the system. However,

$$dU = \sum \varepsilon_i dN_i + \sum N_i d\varepsilon_i$$

and corresponds to

$$dU = dQ_{rev} - pdV$$

which is the first law in macroscopic thermodynamics.

11.7   ENTROPY AND THE SECOND LAW

From Boltzmann's equation where entropy is given by $k \ln W^0$, we can also provide a microscopic interpretation of the entropy. When heat or work is added to a system reversibly, the distribution is not influenced. However, due to either irreversible heat transfer or irreversible work, the orderly distribution gets to be more disorganized or disorderly. Thus, the equation $S = k \ln W$ provides a microscopic relation between disorder and entropy.

At microscopic level, probability of distribution is a measure of disorder. W^0 represents the equilibrium distribution N_i^0; thus if $\ln W^0$ is predominantly large and fluctuation from $\ln W^0$ is vanishingly small, we elucidate the second law that the entropy is a maximum at equilibrium and fluctuation from equilibrium value is negligibly small. Expanding $\ln W$ about the equilibrium value, we write

$$\ln W(N_i) = \ln W^0\left(N_i^0\right) + \left(\frac{\partial \ln W}{\partial N_i}\right)_{N_i^0} \Delta N_i + \frac{1}{2}\left(\frac{\partial^2 \ln W}{\partial N_i^2}\right)_{N_i^0} \Delta N_1^2 + - - -$$

and since $\ln W^0$ is a maximum, the first derivative vanishes.

Thus,

$$\ln W(N_i) - \ln W^0\left(N_i^0\right) = \frac{1}{2}\left(\frac{\partial^2 \ln W}{\partial N_i^2}\right)_{N_i^0} \Delta N_i^2 + - - - -$$

Since

$$\ln W(N_i) = \sum N_i \ln \frac{g_i}{N_i} + \sum N_i$$

and

$$\frac{\partial \ln W}{\partial N_i} = \sum \ln \frac{g_i}{N_i}$$

$$\frac{\partial^2 \ln W}{\partial N_i^2} = -\sum \frac{1}{N_i}$$

Thus

$$\ln \frac{W(N_i)}{W\left(N_i^0\right)} = -\frac{1}{2}\sum \frac{\Delta N_i^2}{N_i} = -\frac{N}{2}\sum \frac{N_i}{N}\left(\frac{\Delta N_i}{N_i}\right)^2$$

Since N_i/N is the probability to find N_i particles in level ε_i, the term inside the summation term is just the moment of $\left(\dfrac{\Delta N_i}{N_i}\right)^2$ (i.e., the averaged value). Therefore,

$$W(N_i) = W\left(N_i^0\right) e^{-\frac{N}{2}\left(\frac{\Delta N_i}{N_i}\right)^2}$$

and for, $N \gg 1$, $W(N_i)$ is negligibly small when ΔN_i is finite. However, for small systems, the fluctuation will be large and it is difficult to define an equilibrium state. The above consideration provides a molecular interpretation of the second law in thermodynamics.

Index

Printed in the United States
by Baker & Taylor Publisher Services